Springer Tracts in Civil Engineering

Springer Tracts in Civil Engineering (STCE) publishes the latest developments in Civil Engineering—quickly, informally and in top quality. The series scope includes monographs, professional books, graduate textbooks and edited volumes, as well as outstanding PhD theses. Its goal is to cover all the main branches of civil engineering, both theoretical and applied, including:

- Construction and Structural Mechanics
- Building Materials
- Concrete, Steel and Timber Structures
- Geotechnical Engineering
- Earthquake Engineering
- Coastal Engineering; Ocean and Offshore Engineering
- Hydraulics, Hydrology and Water Resources Engineering
- Environmental Engineering and Sustainability
- Structural Health and Monitoring
- Surveying and Geographical Information Systems
- Heating, Ventilation and Air Conditioning (HVAC)
- Transportation and Traffic
- Risk Analysis
- Safety and Security

Indexed by Scopus

To submit a proposal or request further information, please contact:
Pierpaolo Riva at Pierpaolo.Riva@springer.com (Europe and Americas) Mengchu Huang at mengchu.huang@springer.com (China)

More information about this series at http://www.springer.com/series/15088

Cecilia Bolognesi · Daniele Villa
Editors

From Building Information Modelling to Mixed Reality

Springer

Editors
Cecilia Bolognesi
Dipartimento ABC
Politecnico di Milano
Milan, Italy

Daniele Villa
Dipartimento di Architettura
e Studi Urbani (DASTU)
Politecnico di Milano
Milan, Italy

ISSN 2366-259X ISSN 2366-2603 (electronic)
Springer Tracts in Civil Engineering
ISBN 978-3-030-49277-9 ISBN 978-3-030-49278-6 (eBook)
https://doi.org/10.1007/978-3-030-49278-6

This Springer imprint is published by the registered company Springer Nature Switzerland AG
The registered company address is: Gewerbestrasse 11, 6330 Cham, Switzerland

Preface

This collection of chapters represents a true cross section on the main themes that our researches have undertaken in the recent years and which, as a common matrix, refer to the theme of representation in a BIM era.

The ability offered by digital means to face the survey of the existing, the development of our skills in the modelling of artefacts or processes, the ability to enrich them semantically according to the different purposes of use has opened our research to infinite applications in which each of us has found its own experimentation.

The occasion for the collection of the chapters presented here starts from the debate developed during the 2nd Brainstorming BIM, a meeting held in Milan in October 2019 after the first edition in 2016: a moment of discussion on the most recent issues that links representation to BIM-oriented modelling.

The research of this second meeting turned out to be fully responsive to a more advanced phase of BIM by refining practices that only a few years earlier focused on modelling itself, as a new cognitive basis of the building essential for any study, starting from its physical dimension.

Many of the papers here presented always use the historical architecture as the main material that gives the possibility of many processing giving the seed for plenty new branches of research still developing. Historical buildings are the product of a process of modification and stratification developed through centuries, and for this reason, they become the occasion of inexhaustible researches; in recent times, it has become possible to hypothesize and develop digital twins of buildings and urban contexts based on a wide array of datasets stemming from several survey techniques which add up to information layers, such as point clouds and semantically enriched construction elements, that are not just a mere representation, rather a proxy of reality with the following possibility to enrich the object attributes with dedicated databases. From this point, the possibility of creating informed models has broadened the horizon of research by demonstrating the possibility of development in the technical field as well as in the communicative or educational one.

There are several papers presented here that refer to the historical building as the basis of the elaborations relating to BIM for cultural heritage; the researches here describe are divided between the analysis of general workflows and the definition of more specific issues, now dedicated to verify the accuracy of the models, now to their ontological and semantic representation or modeling for the dissemination of research results through accessible storytelling techniques. These are dissertations ranging from model reliability levels to the elaboration of virtual processes for the storytelling, to the reworking of spaces designed in other eras and never realized, to the elaboration of HBIM systems for enrichment of cultural heritage management. It is a second season of BIM which therefore provides for its use in an in-depth manner and clearly aimed at specific areas of development.

In the technical field the applied research sector for the development of the AEC takes advantage of all the instruments up to organize totally digital workflows and coordination for the management of the built as well as the construction of the new; on this issue, there is a shorter list of papers focused on management. It is the fourth industrial revolution that has touched in the last decades the construction world where digitalization of industrial processes has implemented the development of platforms that can communicate many information for different users; not only a technological innovation but a more complex one involving organization and management where BIM can expand its application scale as a tool for sharing data.

A last part of the papers further refines research into all sectors related to immersive realities to enhance the knowledge of the area and share the value of heritage, both tangible and intangible, or develop new immersive control systems with a wider audience; in the field of education, it is the era of experience offered by ICT for the creation of serious games or edugames for reaching out wider audiences with educational contents. Although VR and AR have found wide development in the previous chapters, until becoming an essential part of the story, nevertheless it is perhaps in this sector of papers that virtuality has found greater freshness, which has become the object of modelling and subject for entry into new stories and reality. From this field also, the first AI experiments related to modelling move their first steps out.

Milan, Italy

Cecilia Bolognesi
Daniele Villa

Peer Reviewers

Cecilia Bolognesi—Politecnico di Milano
Damiano Aiello—Università degli Studi di Catania
Sonia Lupica—Spagnolo Politecnico di Milano
Graziana D'Agostino—Università degli Studi di Catania
Fausta Fiorillo—Politecnico di Milano
Matteo Lo Giudice—Politecnico di Torino
Claudio Mirarchi—Politecnico di Milano
Fabiana Raco—Università degli studi di Ferrara
Jessica Rumor—Università di Roma La Sapienza

Contents

BIM for Built Cultural Heritage: Semantic Segmentation, Architectural Stratification and LOD of the Baptistery of San Giovanni in Florence

Carlo Bianchini and Giorgia Potestà

Abstract The 3D virtual modeling of any artifact involves the organized composition of digital elements. This process is common to any 3D modeling workflow but, when applied to BIM systems, shows a good correspondence in the construction processes between the virtual and the real world. In fact, any building can be considered as a coordinated set of simple technological elements, linked together by design logics and construction techniques that are quite predictable and controllable. Even existing buildings can be "deconstructed" into recurring elements, especially if the architectural survey operations reveal the regularity of certain geometric patterns; however, parametric and informative modeling of built artifacts is much more difficult, both in terms of geometric transposition of the continuity of the real world and of its qualitative and semantic description. These difficulties are also associated with the intrinsic rigidity of the parametric modeling workflow, subjected to "libraries" of digital objects that clash against the variability and uniqueness of the built environment, especially when "historic" or in a poor state of conservation. While the transition from numeric to geometric models involves a crucial critical "discretization" operation, in the construction of the BIM, the model involves a further interpretative step: the semantic structuring of its compositional elements. In this framework, the presented study becomes an application opportunity for the theoretical issues highlighted above, using the Baptistery of San Giovanni in Florence as a case study. The modeling of this monument, relevant from a historical point of view and for its particular stylistic and geometric features, is based on a massive, integrated survey conducted by the University of San Diego. Specifically, we intend to analyze not only how to discern the constructive elements through their semantic segmentation, but

C. Bianchini (✉) · G. Potestà
Department of History, Representation and Restoration of Architecture, Sapienza University of Rome, Piazza Borghese 9, Rome, Italy
e-mail: carlo.bianchini@uniroma1.it

G. Potestà
e-mail: giorgia.potesta@uniroma1.it

C. Bolognesi and D. Villa (eds.), *From Building Information Modelling to Mixed Reality*, Springer Tracts in Civil Engineering,
https://doi.org/10.1007/978-3-030-49278-6_1

above all how to connect them and make them interact with each other. As a result, we aim at providing a representation of the different evolutionary phases and above all propose a more effective meaning for the Levels of Development parameter when applied to built heritage.

1 Introduction

The evolution of digital technologies at the service of the representation of architecture has allowed the achievement of objectives previously unthinkable for the survey, knowledge and communication of the built heritage. By integrating the possibilities offered by three-dimensional, solid and parametric modeling tools, with those of data management and sharing systems, BIM systems—Building Information Modeling—promise new scenarios for storing and managing large amounts of information for the knowledge of cultural heritage (Quattrini et al. 2016; Lopez et al. 2018). The research intends to analyze the potential and the critical aspects deriving from the integration of BIM systems in the documentation, knowledge, communication and management processes of the Cultural Heritage.

The potential of BIM systems oriented to the so-called Heritage-BIM (H-BIM) is undeniable; however, the theoretical and operational questions relating to the rigidity of the modeling tools remain unresolved when elements of the cultural heritage are involved, where in fact high flexibility is required (Hichri et al. 2013; Inzerillo et al. 2016). The starting point is the three dimensional survey conducted by integrating non-contact methodologies, able to allow the processing of various types of models (3D and 2D, numerical, geometric and texturized) whose relationship with the HBIM platforms is to be studied. Although much still needs to be clarified, the main characteristics of the final product can already be indicated. First of all, it must respect all the potentials of BIM systems: interaction, interoperability, simulation of construction processes (from construction to maintenance). Furthermore, it must be able to host and display 1D (texts) and 2D (drawings, images) information used to build it (Bianchini and Nicastro 2018). So it should interact with 3D data coming from instrumental investigations. Last but not least, it should provide proof of the reliability of the digital objects that make up the model, according to the already known concept of the Level of Reliability (Bianchini and Nicastro 2018). The methodology adopted by the project is fairly consolidated. Starting from a systematic review of the existing literature (Lopez et al. 2018; Hichri et al. 2013), the H-BIM application will try to answer the previous questions by outlining strategies that can lead to validated workflows. The Baptistery of San Giovanni in Florence will offer excellent application ideas, being a historicized monument, rich in architectural stratifications and subsequent restorations, for which it will be interesting to understand not only how to discern the construction elements but above all how to tie them and make them interact with each other.

1.1 *The Research Process*

The research workflow has entailed different steps: a survey phase, which envisages an examination of the current regulations in the field of Building Information Modeling, a substantial part concerning the historical and archival investigations on the case study aimed at the realization of a database of information from which to draw for modelling; and finally the data acquisition phase through a non-destructive survey campaign and the normalization of the acquired data. Central role is given to "knowledge", as a means for evaluating the value of the tangible and intangible aspects of the building and for structuring a digital environment suitable for storing, modeling, managing and querying data. Finally, a last phase has been referring to semantic segmentation that involves the decomposition of the artifact into its constructive elements. This process helps us to understand which parts make up the architecture and how they relate one to another. In a completely consequential way, then, the distinction between these elements takes place, in families of replicable or non-replicable objects and according to their variant and invariant attributes following the logic of the parametric BIM modeling. The reconstruction of the architecture just decomposed into a 3D BIM model of the building is the final outcome. This model shows different Levels of Detail and the possibility of displaying additional information, such as the different building phases to which the modeled elements belong or their detail levels.

2 About the Monument

In dealing with the case study from the point of view of 3D modeling, and in particular in BIM modeling, historical, archival and documentary research are of substantial importance for the knowledge of the building. Specifically, in-depth research was conducted from the point of view of historical knowledge and territorial framing of the main building phases, archive documents and previous surveys, useful for understanding the construction of the monument. The investigations were conducted directly in the city of Florence, within the main libraries of the city, and in particular at the Uffizi Library (ex-Magliabechiana library) and at the Instituto Nazionale di Studi sul Rinascimento. We have also taken into account the recent publication of the Proceedings of the International Study on the Baptistery of San Giovanni held in Florence in November 2014.

2.1 *The Construction of the Monument*

The Florentine Baptistery is a regular octagonal masonry construction, characterized by alternating sides with portals and sides without openings to the first interrupted register on the south side, where there is the quadrangular apse, called scarsella, built in the 13th century. The dome has an octagonal plan covering, invisible from the outside by means of a pyramid-shaped structure with flat layers covered with white marble slabs. From a structural point of view, as evidenced by the studies of Bartoli, Betti and Monichetti (Gurrieri 2014), two levels can be recognized in height: one from the height of the floor to the soffit of the matroneo, and the other from the extrados of the matroneo to the gallery. The path of the matroneo represents the connection between the two levels, defining a ring that runs along the perimeter, interrupted only by the scarsella; a second continuous annular element along the eight faces is present at the top of the second level, constituting the support of the dome and the drum.

2.2 The Main Construction Phases

Until the second half of the sixteenth century, we cannot speak of a question about the dating of the Baptistery of San Giovanni, as contemporaries took the Florentine tradition to the truth that the monument was born in Roman times as the ancient Temple of Mars. Among these, we can mention Dante, Giovanni Boccaccio, Giovanni Villani, Francesco Sacchetti, Aretino, Poliziano, Leonardo Bruni and Vincenzio Borghini (1580). Girolamo Mei, who, first, proposed a Lombard origin, thus starting the debate that has continued up to our days, rejected this thesis. Today the positions of scholars are contrary to the idea of a Baptistery first as the Temple of Mars, a hypothesis abandoned by all but by Degl'Innocenti (2017), supported by Carlo Cresti, and most of them set the Foundation of the monument at the Medieval Period.

As mentioned above, the analysis of the monument's construction phases is preparatory to understanding this, as well as the possibility of inserting this information into the BIM platform for each element modelled using "phase filters".

Three main phases have been recognized in the construction of the monument (Paolucci 1994): A first phase, defined as Roman, at the end of the 11th century, which consists in the construction of the structure without octagonal-shaped cladding, a double stone foundation ring, a constructive system with vertical pylons and radial walls, the provision of a segmented dome and the cladding marble only with regard to the first internal register; the exterior remained with a visible stone finish or perhaps plastered. A second phase, dated between the end of the eleventh century and the beginning of the 12th, defined as "marmoreal", is characterized by the realization of the marble covering of the first two external registers, the shielding of the women's gallery with two marble mullioned windows and the realization of the lantern; the roof becomes pyramidal and covered with marble slabs. The third phase, dated to the first half of the thirteenth century, defined as "gothic", is characterized by the realization of the marble covering of the interior of the second register, the full covering of the covering surfaces and the matroneo through the mosaic with a gold background and the external covering of the cantonal marble pylons; finally the quadrangular scarsella is made. Further work on the monument involves the construction of bronze doors, sculptures placed above them and interior furnishings. In the XVIII century the archaeological excavations led by Edoardo Galli began, which led to the discovery of the Roman Domus below the foundations of the Baptistery.

Further work on the monument involves the construction of bronze doors, sculptures placed above them and interior furnishings. In the XVIII century the archaeological excavations led by Edoardo Galli began, which led to the discovery of the Roman Domus below the foundations of the Baptistery.

3 Acquisition Campaign

The survey on which the proposed study is based was conducted in 2013 by Center of Interdisciplinary Science for Art, Architecture, and Archaeology (CISA3) at the University of California San Diego. Different data acquisition methods were used, including terrestrial laser scanning (TLS), Structure from Motion (SfM) photogrammetry, infrared therography, stereo spherical gigapixel imaging, and ground penetrating radar. Data were fused using CISA3's proprietary point cloud rendering software.

A total of 62 individual TLS scans were performed (Fig. 1), and aligned to create an overall point cloud with a half billion 3D data points acquired (Degl'Innocenti 2017). The large amount of data acquired resulted in not only a more faithful representation of the monument, but enable researchers to go beyond the wall surface, defining information regarding the materials and their state of preservation; this is all information that can be integrated into the BIM model and that determines the level of objective reliability of each modeled element (Fig. 2).

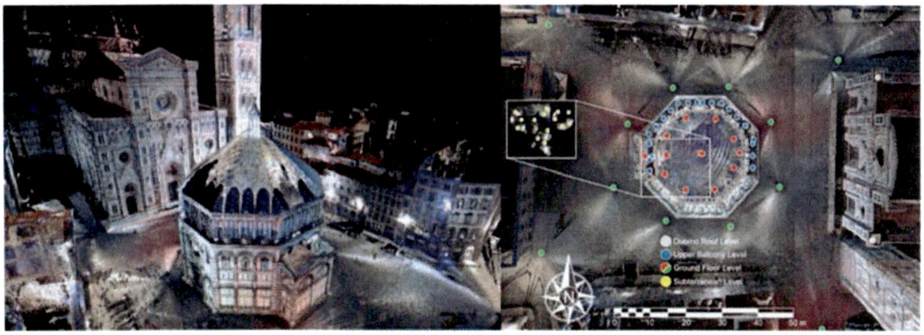

Fig. 1 View of the point cloud with detail of the scan positions. Elaboration by Hess (2018). Software viscore

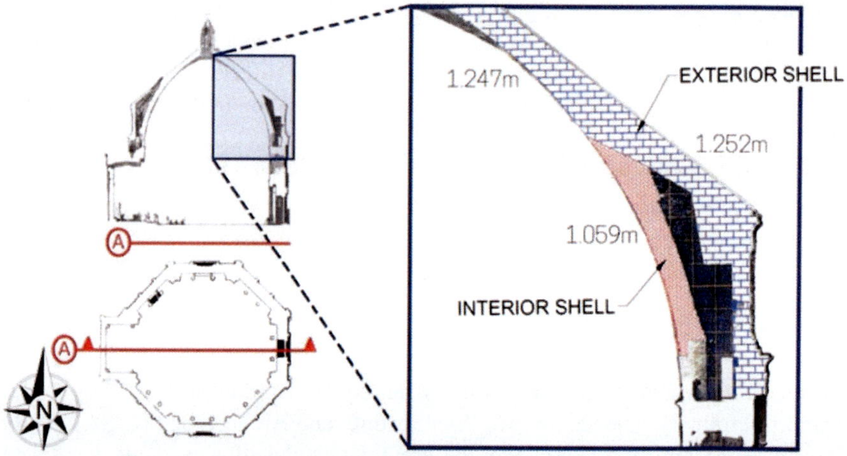

Fig. 2 Visualization and analysis of laser scanning data. Construction detail of the dome cap. Elaboration by Michael Hess. From viscore to CAD

4 Semantic Segmentation

As already mentioned, each artifact can be broken down into a coordinated set of simple and complex elements, which together determine its structure. The analysis and decomposition of these parts within a built architecture is called, in a more or less appropriate manner, Semantic Segmentation (Fallavolita et al. 2015).

In reality, semantic segmentation is a deep learning algorithm that associates a label or a category with each pixel of an image. It is used to recognize a series of pixels that form distinct categories. Segmentation is used in numerous applications such as autonomous guidance, medical imaging and industrial inspection and in recent years has also been applied to the field of architectural survey.

The segmentation operation facilitates the study of historical monuments and integrates heterogeneous information and attributes, useful for describing and characterizing the object in question. We can segment and organize any model as a system of cognitive architecture. The building can be described as a collection of hierarchically organized elements, identified by a specific architectural vocabulary. The segmentation of the model into sub-elements can be based on the analysis and composition of the formal structure of architectural objects, organized in hierarchical levels and aggregation classifications. Particular attention in this operation must be placed in analyzing how these objects are assembled from a typological and morphological point of view, since this problem arises every time one approaches the BIM modeling of an artifact: understanding how each element is constituted and how binds to others within the model. Subdividing the artifact into hierarchical elements helps the BIM modeling in which the model is divided into its component parts, based on libraries of objects, morphological elements and/or different geometries.

The segmentation of historic buildings also requires an architectural vocabulary correlated with the knowledge of all those structural, functional and decorative components, such as moldings, architectural orders, trabeations and profiles, and their geometric and topological characteristics, in terms of features and connections between the parts, in order to completely describe the building.

In the field of HBIM, the heterogeneity of data, information and terminology often makes hybrid solutions necessary. The classification of architectural forms on the basis of dimensional and geometric parameters, but also semantics, can help the recognition, interpretation and classification of the characters of historical buildings.

As for the case study, the semantic segmentation was made on the basis of ortho-images (Fig. 3), which are therefore two-dimensional models, and concerned not only the decomposition into architectural elements according to different levels, but also the hierarchical identification of architectural elements, such as openings, supports and capitals (Fig. 4).

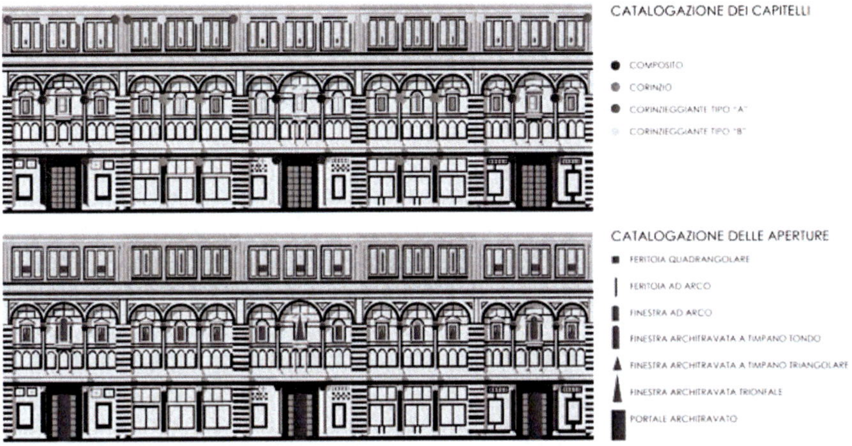

Fig. 3 Hierarchical identification of architectural elements. Elaboration by author on ortophoto

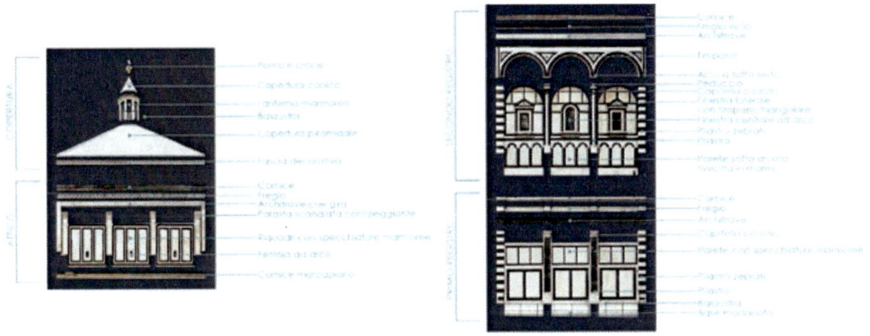

Fig. 4 Semantic segmentation of the monument. Elaboration by author on ortophoto

5 Regarding LODs—Levels of Development

The BIM and its implementation in the projects make it increasingly important to clarify the concept of graphic detail of the projects and LOD (Rossi and Palmieri 2019) present in the BIM models. On several occasions, it has become evident that there is confusion between the designers and the realities involved at various levels in the BIM design on the meaning of LOD.

For each phase of the project, considering the new constructions, it is of fundamental importance that the level of detail required, both for graphic and geometric information and for attributes, is well defined. A great deal of information can converge in the BIM model, and it is therefore important that those useful to the client and used by the various professionals who collaborate in the project reside in the model, in the different phases of the design. So the level of detail of the BIM model increases as the project proceeds: in the first phase the information that characterizes the model is usually that relating to the existing situation, while in the subsequent phases we pass from a simple concept model to a virtual model ("as-built") made real due to a detailed and, in fact, constructive level. The design practice effectively involves the development of information and geometries at different speeds and these aspects can come from the various collaborators on the project team. In the same way, when we consider an existing building, especially if it belongs to the sphere of the built historical heritage, we are faced with the same problem, that is to define in advance the different levels of development of the project, in which the starting level is already the "as-built", or rather "as-is": the current situation is the historical building in front of us, characterized by a whole series of attributes, both geometric and computerized, more or less detailed.

The LODs have the task of precisely defining the level of detail of various types of information contained within the model. The term LOD represents the "Definition level" (or "Development level") that the objects present in the BIM model must take in the various phases of the design. As defined in the initial phases of a BIM design, the client draws up a document named EIR (Employer's Information Requirements) containing fundamental requirements to be included in the design.[1] The definition of the characteristics of each individual LOD is a theme addressed by two important normative references, one of American origin and one Italian (Fig. 5 http://etwinning.indire.it/).

In the American context, the American Institute of Architects (AIA) has published an LOD framework for the AIA Protocol G202-2013 Building Information Modeling, where the term LOD refers to the level of development needed in relation to the contents of the elements of the model; the choice to use the definition "level of development" instead of "level of detail" is motivated by the fact that an element, although it may appear visually detailed, could actually be generic. According to this document, the degree of development achievable through the drafting of a building model defined through a BIM approach is divided into five levels arranged by hundreds, from the LOD 100 in which the element is represented in a generic manner, to the

[1]AEC (UK) BIM protocols.

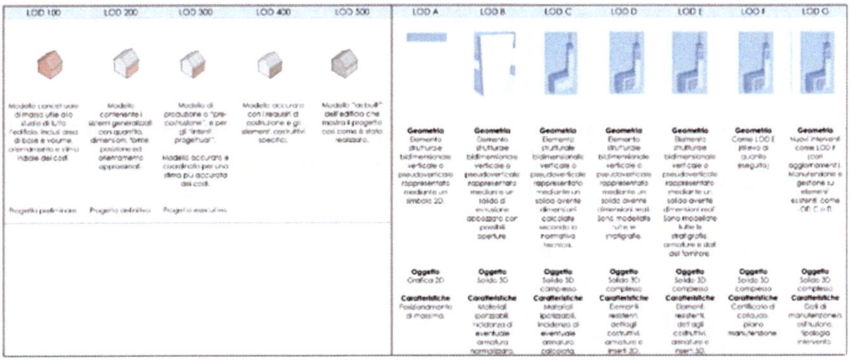

Fig. 5 Classification of LOD reported in the main international guidelines: American and Italian classification

LOD 500 where the element faithfully reflects reality and is a representation verified on site in terms of shape, position, size, etc.

The Italian legislation, on the other hand, STANDARD UNI 11337-4: 2017,[2] provides for an articulation of LOD in alphabetical sequence, where for restoration interventions the LODs from A to E are not significant except for some opportune simplifications of the specialized model. On the contrary appear important the LOD F, where the objects express the virtualization detected on the site of the specific existing system (as built) and are defined for each individual product the management interventions, maintenance and/or repair to be performed over a programming period, and the LOD G, which concerns the historicized representation of the flow of the useful life of a specific system updated with respect to what was treated or installed in a previous intervention. The quantitative and qualitative characteristics (size, shape, location, orientation, etc.) are updated with respect to a previous state of affairs. The Italian legislation has finally defined the possibility of integrating the attributes of the LODs of some objects even with 2D nodes and views, and the method of integrating the LODs of the objects to the internal of the overall digital information process.

5.1 Level of Development Versus Level of Detail

Modern BIM authoring software is able to model construction elements with different graphic details, from a schematic display up to a particularly high level of detail. But from what is possible to see inside the architectural BIM software, there seems to be a close analogy between the level of development and the level of detail of the model components. It is therefore easy to think that it is sufficient to increase the detail in order to be able to pass from a LOD to another, and that the Level

[2] UNI Norma Nazionale 11,337–4: 2017. Edilizia e opere di ingegneria civile—Gestione digitale dei processi informativi delle costruzioni, Milano 2017.

of Development is equivalent to the graphic detail. The reality is quite different: for example, by modeling a building package we can define progressively different LODs that concern the level of development of the element and its characteristic, different for geometric components. Here then is that the LOD differs from the graphic detail for the attributes that can populate the BIM objects according to the design phase, which is the LOI (Level of Information). One cannot think of a BIM object that does not have a series of non-graphical information added to it (the info attributes) in its higher LODs. Precisely this characteristic makes the sequence of LODs unique and not to be confused with the purely graphic detail, Level of Detail that is applied to the only geometric component of the modelled element.

5.2 *Firmitas, Utilitas E Venustas in the Modeling of Levels of Detail*

Having clarified this doubt, we then focused on the analysis of the concept of "parametric" and its affinity with the concept of "architectural module" and modularity. The parametric term refers to the relationships existing between all the elements of the model, which allow for the coordination and the management of the changes. In mathematics and in mechanical design CAD systems, the numbers or characteristics that define this type of relationship are called parameters, hence the use of "parametric" in reference to the operation performed by the software; As a result of this coordination, the changes made at one point are extended to the entire Revit project. Since our task is to parameterize as much as possible a historical architecture within the BIM platform, we have found the need to find an affinity between the parametrization and the concept of module in the architectural order, which turned out to be completely natural. In the concept of architectural order, in fact, since antiquity, the concept of series, succession, repetition, in the sense of the whole, of instruments suitable for reading an architectural organism as a numerical and harmonic system, continually re-proposable through a module.

In De Architectura book III (2002), Vitruvius dwells on the need for the composition of the temple to be based on proportional schemes, that is, on the commensurability of each individual member of the work and of all the members of the work as a whole, by means of a determined unit of measure or commodulatio; this for Vitruvius corresponds to the diameter of the column measured at the imoscapo.

Furthermore, the Vitruvian procedure of subsequent partitions can be applied to orders distinguishing three main levels, intimately related to the Vitruvian triad of firmitas, utilitas and venustas, as theorized by Riccardo Migliari (1991): the Constructive Level, that of the large masses and of the first proportioning; the Functional Level, which includes the smaller parts, generated by division of the other elements. In this level, the parts are characterized by having specialized functions, which justifies the name given to their grouping; and finally the Decorative Level, in which each element is completely modelled by means of the juxtaposition and proportioning of

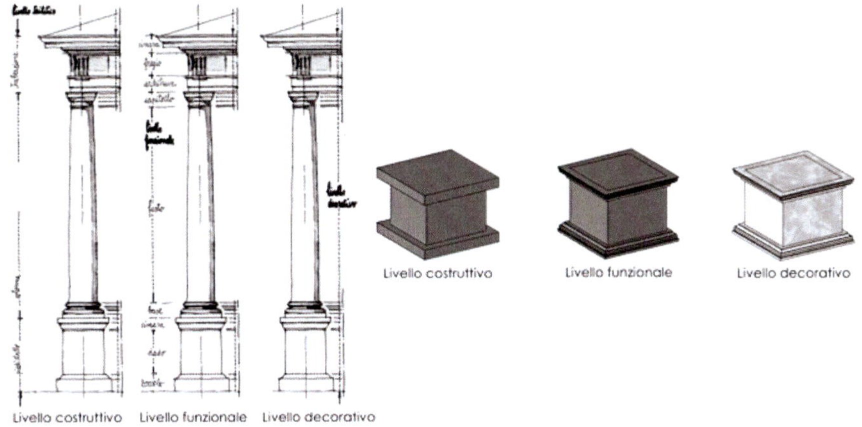

Fig. 6 The three levels of detail in the construction of the Doric order of Riccardo Migliari compared to modelling levels of detail of a pedestal

the mouldings. It is here that every order is characterized. Is it therefore possible to apply this same subdivision in levels, based on the Vitruvian logic, to the decomposition of the Levels of Detail? We know that inside the BIM authoring Revit it is already possible to view the elements according to three different levels of detail, which are called Course, Medium and High, without any scientific reflection at the base (Fig. 6). In fact, as it is possible to verify if we model a classical architectural order within the platform by changing the visualization in the different levels of graphic detail, they appear congruent with the three aforementioned levels: a first modeling for large masses, one in which all the functional elements are present, and finally that which describes the complete element of its decorative parts.

This model, characterized by the display of different degrees of graphic detail, as well as the Level of Reliability (the degree of objective reliability of the individual elements modelled) (Bianchini and Nicastro 2018), or the Level of evolution, in which emerges the representation of the different construction phases of the model, can to be just some of those we have called Thematic Levels (Fig. 7). These are levels applicable to each H-BIM model depending on the objectives of the model itself and what is to be communicated.

6 BIM Modeling

With respect to the modeling of the monument inside the Revit platform, it was essential not only to break down the architecture, as we have seen, in different parts and to investigate its connections, but to understand which of these elements were repeatable and therefore modeled to internal of parametric families or system families (Murphy et al. 2013), and which instead had to be instances modeled in place or as

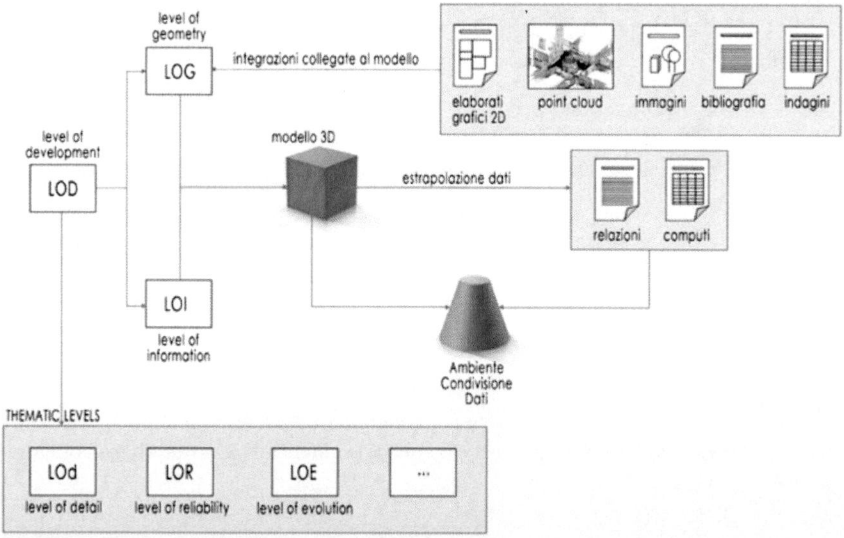

Fig. 7 Ri—definition of data flow in digital process through the LOD. Elaboration by author

a imported mesh models; a phase, so to speak, of meta-modeling, but which made it possible to shed light on the process of modeling historical architecture.

The point clouds acquired by Faro laser scanner were recorded and aligned within Scene and imported in Viscore software, from which it was possible to obtain a series of high-resolution shapshoots that are pre-technical for modeling the detailed elements; from this we then moved into Autodesk Recap to help import the cloud recorded in Revit. A fundamental operation within Recap was to group the scans, dividing them by acquiring the exterior, the interior and the subterreans, and above all the creation of specific Scanregions for the dome, the different registers and the scarsella in such a way as to facilitating direct modeling from the cloud in Revit, since the BIM platform, in the cloud import phase, also allows the importation of any segmentations made in Recap.

Once the point cloud was imported into Revit, we proceeded with the geolocation of the project and the definition of the orientation, aligning the project north with the real north and thus generating views and elevations in line with the definitions used to date. We then proceeded with the setting of reference planes, essential not only to generate views in plan or section, but above all to constrain the objects modeled to specific plans or alignments of the architecture itself (Fig. 8).

We have chosen to start modeling the building envelope as a mass, directly drawing from the imported point cloud. This operation allows us, once the general envelope is modeled, to apply system families such as walls and floors directly to the mass by selecting the walls and reference levels of which you want to extrude the floor, proceeding then to define the stratigraphy as a subsequent step. In the same way it is possible to apply to the mass the Continuous façade System that facilitates the

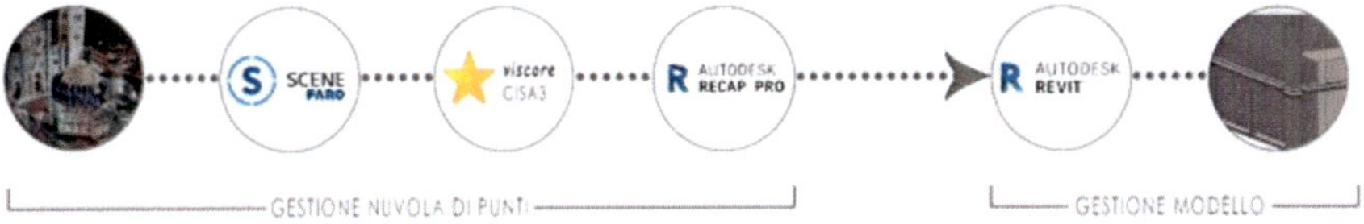

Fig. 8 Workflow through the softwares

creation of masonry openings, such as windows or glass windows and to be able to draw the grid for dividing them. In this phase everything is modeled directly based on the point cloud. The stratigraphy of the walls was defined on the basis of the recent studies and structural analyses that concerned the monument and which are traced in the bibliography.

All repeatable architectural elements, (conveniently modeled as such), including windows, columns and some molded profiles (Apollonio et al. 2013), have been treated as loadable families, or as separately modeled elements that are subsequently loaded into the project. The latter were modeled on a CAD basis, importing the pofiles in Revit, or on a detailed orthophoto base captured from the point cloud and directly drawn in Revit (Fig. 9).

Like most parametric modeling software, Revit is also not able to extract data from the points of interest of the cloud in order to guarantee a faithful modeling of the geometries, but as with 2D rendering, even in the BIM modeling the cloud of points is treated as a database, an image to be re-draw and from which to obtain the geometry of the building, and used as a metric trace. This process is clearly performed manually and consequently implies a strong degree of subjectivity in the interpretation and execution of modeling operations.

Fig. 9 Modeling of the Baptistery facades starting from the mass

Finally, a further aspect of the model was investigated, namely that of the possibility of inserting element construction filters, the so-called phase filters. These allow us, in the modeling and management phase of the modeled element, to insert information about the creation phase of the element, providing information about the era of creation and to which building phase the monument belongs (Fig. 9). The phase filters therefore give the possibility of generating an architectural model that also contains a synchronic analysis of the artifact, or that in its "as-is" state it gives an account of the dating of some of its parts, visualizing them simultaneously. Obviously, the platform also makes it possible, as well as for the modeling of new buildings, to generate different views for each phase filter, thus giving the possibility of having a diachronic analysis, or of displaying separately the elements that belong separately to each construction phase.

7 Conclusion

The research, part of a broader project that concerns the collaboration between the two Universities, in Rome and San Diego, has involved the investigation of the different aspects of the BIM modeling of a historical artifact, trying to define a possible operational process and going beyond the obvious limits that this type of modeling constitutes when applied to the historical heritage built.

Clearly another limitation for the modeling of historical buildings, which starts from the acquisition of relevant data, is in the process of converting file extensions to be imported within the BIM platform. In order for this to show itself as an effectively interoperable and data sharing tool also for the historical heritage built, a greater degree of freedom in receiving data with different extensions is indispensable in order to simplify the import operations.

The growing use of these means of representation for the projects concerning the field of architecture, engineering and now also restoration, requires us, as scholars, to find operative methods for the application in the cultural field without forgetting the purpose that such modeling arises. In fact, we must not forget that BIM is a digital information sharing process, and therefore, as we are reminded by the Italian legislation of which we spoke earlier, not everything must necessarily be accurately modeled if technically unsustainable and integrated with attributes from external objects.

Acknowledgements We thank the University of California San Diego for the hospitality in their laboratories and the sharing of the relevant data of the Baptistery and the analyses carried out. Special thanks to prof. Dominique Rissolo and Falko Kuester, to Vid Petrovic, Michael Hess and Eric Lo.

References

Apollonio FI, Gaiani M, Sun Z (2013) 3D modeling and data enrichment in digital reconstruction of architectural heritage. In: International archives of the photogrammetry, remote sensing and spatial information sciences, Vol XL-5/W2 XXIV international CIPA symposium, 2–6 Sept 2013. Strasbourg, France

Bianchini C, Nicastro S. From BIM to H-BIM. In: 2018 3rd digital heritage international congress (DigitalHERITAGE) held jointly with 2018 24th international conference on virtual systems & multimedia (VSMM 2018). IEEE, pp 1–4

Bianchini C, Nicastro S (2018) La definizione del level of reliability: un contributo alla trasparenza dei processi historic—BIM. In: Empler T (ed) 3D modeling & BIM, applicazioni e possibili sviluppi futuri", (a cura di), Facoltà di Architettura Sapienza Università di Roma, 18–19 Apr 2018. Pp 209– 25

Bossalino F (2002) De Architettura libri X, Roma

Degl'Innocenti P (2017) L'architettura del Battistero fiorentino: progetto, appalto, costruzione, vicende, Firenze

Fallavolita F, Ballabeni M, Foschi R, Perugini G (2015) Semantic description of the three-dimensional models of Bologna porches. In: SCIRES-IT scientific research and information technology ricerca scientifica e tecnologie dell'informazione, vol 5, Issue 1, pp 31–40

Gurrieri F (2014) Il battistero di San Giovanni: conoscenza, diagnostica, conservazione. In: Atti del convegno internazionale Firenze, Firenze, 24–25 Nov 2014

Hess M (2018) Diagnostic structural health assessment through layered integration of non-destructive imaging data. A dissertation submitted in partial satisfaction of the requirements for the degree doctor of philosophy in structural engineering, University of California San Diego

Hichri N et al (2013) Review of the "As-built bim" approaches. In: Proceedings of the 3D-ARCH-3D virtual reconstruction and visualization of complex architectures. international society for photogrammetry and remote sensing, Trento, Italy, pp 107–112

Inzerillo L, Lo Turco M, Parrinello S, Santagati C, Valenti GM (2016) BIM e beni architettonici: verso una metodologia operativa per la conoscenza e la gestione del patrimonio culturale (BIM and architectural heritage: towards an operational methodology for the knowledge and the management of Cultural Heritage. Disegnarecon 9(16):16.1–16.9

Lopez FJ, Lerones PM, Llamas J, Gomez-Garcia-Bermejo J, Zalama E (2018) A review of heritage building information modeling (H-BIM). Multimodal Technol Interact 2(2):21

Migliari R (1991) Il disegno degli ordini e il rilievo dell'architettura classica: Cinque Pezzi Facili. In: Disegnare Idee Immagini, anno II, vol 2. Roma

Murphy M, Mcgovern E, Pavia S (2013) Historic building information modelling—adding intelligence to laser and image based surveys of European classical architecture. ISPRS J Photogram Remote Sens 76:89–102

Paolucci A (1994) Il battistero di San Giovanni a Firenze, Firenze

Quattrini R, Clini P, Nespeca R, Ruggeri L (2016) Misura e historical information building: sfide e opportunità nella rappresentazione di contenuti 3D semanticamente strutturati. in Disegnare Con, Le dimensioni del BIM, vol 9, no 16. Roma

Rossi A, Palmieri U (2019) LOD per il patrimonio architettonico: la modellazione BIM per la fabbrica di Solimene. Disegno (4):213–224

Scan-to-BIM Process Versus 3D Procedural Modelling of Gothic Masonry Vaults

Vincenzo Bagnolo and Raffaele Argiolas

Abstract The so-called "Scan to BIM" process has been widely utilized for many years now: deriving 3D models from point clouds often a local modelling of geometric components is necessary. This leads in most cases to use external modelling tools and or complex local modelling processes involving often procedures quite time-consuming. The paper describes a workflow aimed at the 3D procedural modelling of a quadripartite rib vault system. Starting from feature extraction of TLS data based on geometric assets, we provide a workflow to turn TLS point clouds into the geometric primitives of the complex vaulted systems covering the main body of the church of Nostra Signora della Speranza in Cagliari (Italy). Vaulted systems are architectural valuable elements characterized by a fairly large and varied number of case studies; today, also thanks to the use of TLS systems which we now have available, we can consider the geometric structure with a precision previously, until not long ago, unreachable. The geometrical information of the scanned element is the basis of a procedural modelling that aims at writing an algorithm for the parametrization of the modelling process of the different surface elements of the vaulted systems. The proposed methodology lends itself to an effective application in the BIM environment, in particular in those cases where recourse to local modelling is necessary. Thanks to the parametrization of the process, we can overcome some of the limitations of local modelling, often in contradiction with the BIM philosophy itself, which requires the use of flexible and reusable parametric models.

Keywords Scan-to-BIM · TLS data · HBIM · Cultural heritage · Algorithmic modelling · Masonry cross vaults

V. Bagnolo (✉) · R. Argiolas
DICAAR, University of Cagliari, Cagliari, Italy
e-mail: vbagnolo@unica.it; bagnolo@gmail.com

R. Argiolas
e-mail: raffaele.argiolas@hotmail.com

C. Bolognesi and D. Villa (eds.), *From Building Information Modelling to Mixed Reality*, Springer Tracts in Civil Engineering,
https://doi.org/10.1007/978-3-030-49278-6_2

1 Introduction

Data acquisition, processing and management have always been fundamental topics in cultural heritage documentation and preservation, taking on particular renewed relevance and significance with the advent of innovative surveying methodologies and techniques for several decades now. Laser scanning technology has had wide application for quick three-dimensional data acquisition, revealing itself particularly effective in the survey of complex elements. The use of Terrestrial Laser Scanning (TLS) provides accurate 3D data and high-resolution mapping for as-built modelling of architectural heritage. Using TLS data for estimating geometric properties of architectural elements, the phases of data segmentation and feature extraction become two fundamental steps in the point-cloud processing phase. Without these two preparatory stages, the point cloud itself cannot provide explicit information about the geometric assets, but only 3D positional and intensity information on each point. TLS data are an excellent solution for accurate 3D measurement guided by the aim of revealing surface-based characteristics of complex constructive elements. Starting from the point cloud for geometry extraction and model validation, procedural modelling optimizes the entire process and improves the quality of the results. In the study of historical architecture, local modelling is often required: Scan-to-BIM processes involve the conversion of spatial data from laser scanning (or digital photogrammetry) into parametric applications.

Where the complexity of the elements studied does not allow the creation of parametric objects, an alternative solution is that of procedural modelling. The use of a suite of procedural rules enables us to focus and define geometrical content of local 3D models through an algorithmic modelling process, reusable in a flexible way in the logic of the definition of an HBIM targeted library. The procedural modelling is set to obtain different surface elements starting from TLS data. In HBIM workflows, in most cases, 3D modelling involves the extraction of geometric components from 3D point clouds data, it is not an immediate process; it can often be a laborious and a time-consuming task. Developing an algorithmic modelling for the creation of 3D digital model of intrados surface, the geometric genesis of simple and complex vaulted systems is the basis of the modelling process. To prove the effectiveness of the algorithm and to demonstrate compliance with real case studies, the research must necessarily involve the survey of some vaulted systems. Considering the broad variety of existing systems and the wide range of three-dimensional surface geometries, as well as their different possible arrangements, requires identifying and classify geometries considering the comparison between theoretical models and real case studies. For these reasons, it was decided to concentrate the first phases of the research on the survey and modelling of the surfaces of some quadripartite rib vaults. Regardless of the degree of complexity found in this category of vaulted systems, it was considered appropriate to start from the study of the simple continuous surface elements that make up the complex vaulted system as a whole, before moving on to the articulation of the different possible combinations. The result data is a NURBS local model generated individually for each surface element that, unlike other local

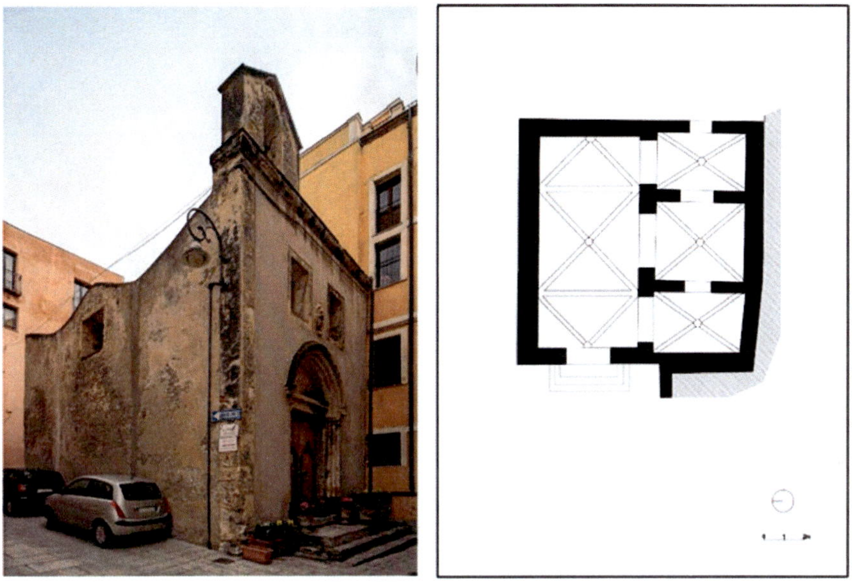

Fig. 1 The church of nostra signora della speranza in cagliari (Italy)

modelling processes, has been reached after a flexible process, reusable for analogous elements by introducing new parameter values, using the same algorithm it is possible to obtain specific local models of different single objects belonging to the same "family". The NURBS model resulting by entering the TLS data in the algorithm was then compared with the point-cloud 3D model generated from laser scanning: the quality of the results demonstrates the effectiveness of the coded algorithm. Here we present the results of a study on the surfaces of the ribbed vault system in the main body of the church of Nostra Signora della Speranza in Cagliari (Italy) (Fig. 1).

Built between the end of the fifteenth century and the middle of the sixteenth century, this small church stands in the historic district of Castello. The church presents a main rectangular space on which three chapels open on its right side. The coverage of the worship space of the church is resolved by adopting a rather particular cross vaults system solution constituted from a sequence of three vaults. A square-plan ribbed cross vault, placed in the central part of the aula, and two half cross vaults arranged in correspondence of the two short sides of the rectangular plan of the main body of the church (Figs. 2 and 3).

The same solution that shows the use of half cross vaults is found inside two other churches in Cagliari also dated back to the first half of the sixteenth century: that of the vault of one of the chapels of the church of Santa Lucia (Bagnolo and Argiolas 2019a), and that of the vault of the sacristy of the Beneficiati in the cathedral of Cagliari.

Fig. 2 3D point cloud laser scan surveys. In red the system of masonry cross vaults under consideration

2 Geometry Detection for Algorithmic Modelling from TLS Da-Ta for BIM

A proper development of a BIM workflow requires a data-rich object-based 3D parametric model holding both geometric and semantic information. In heritage building information modelling (HBIM), starting from surface surveying, data acquisition aims first at defining the geometrical content of 3D models. In most cases, modelling of complex historic elements may not always be developed with automated processing and handling, although this is highly desirable in view of a constant monitoring of the modifications of historical artefacts.

Modelling objects as they are in reality can sometimes be a difficult process that requires manual approaches which means having time-consuming processes that needs skills and gives subjective results that sometimes do not pursue the logic of object definition in BIM environment (Tang et al. 2010).

Scan-to-BIM processes, currently widely used for existing buildings, develops a workflow normally involving laser scanning survey activities, 3D point clouds generation, semantic interpretation and subsequent BIM modelling. The recognition of geometric characteristics and the selection of both geometric and semantic information is a very delicate point in HBIM processes, involving the selection and

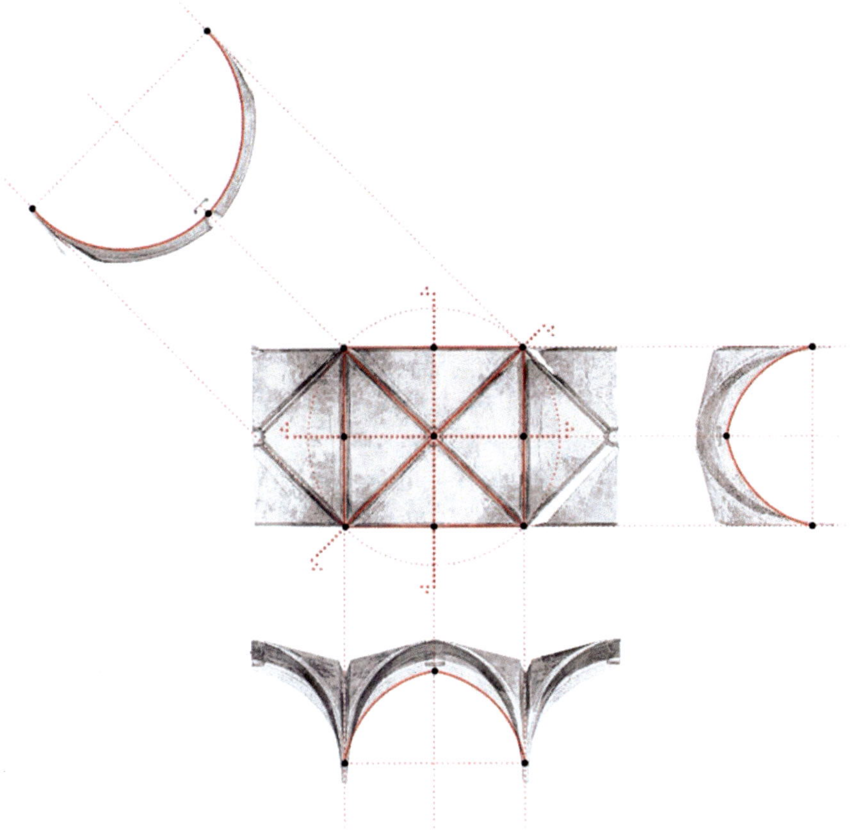

Fig. 3 Sections and profiles of the rib vault system of the main body of the church

addition of the semantic information without affecting the value of the BIM process (Thomson and Boehm 2015). Although in general it is not absolutely necessary to have a geometry modelling strictly adherent to reality (Runne et al. 2001), usually the geometry generation from point clouds for BIM it is essential to obtain a model actually corresponding to the existing building without affecting the value of the BIM workflow. Considering complex artefacts, sometimes it may be necessary to resort the use of local non-parametric models clashing with BIM's standardized procedures Increasing effectiveness by extracting of two-dimensional profiles from the point cloud becomes a fundamental step in the recognition and modelling of geometric features. In addition, we need to focus on historical building treatises and manuals resuming the rules that govern the geometry of historical artefacts between theory and construction practices related to the specific historical design and building culture (Capone 2014).

The historical architectural treatise is of fundamental importance for the analysis of the relationships existing between theoretical principle and constructive praxis, that is among geometry and architecture (Vitali and Natta 2019). The architecture of the groined rib Vaults in Sardinia takes on a regional dimension that refers to the stylistic and constructive tradition existing in the architecture of the Corona d'Aragona in a period comprised between the fourteenth and seventeenth centuries. More generally, the late-gothic architecture of Sardinia constitutes a declination of the so-called "Mediterranean Gothic" characterized by its own defined style next to the Gothic of central and northern Europe.

The studies conducted by Giovanni Curioni regarding the measurement of the surfaces of the vaults and presented in the book "*Geometria pratica*" (Curioni 1869), define a useful synthesis of connection between the purely theoretical issues and the architectural practice, moving from geometric genesis to mathematical description. Curioni examines the cross vaults considering the individual modules consisting of the secondary vaults, which in our case correspond to the surfaces of the caps between the ribs. Depending on the geometry, Curioni distinguishes the secondary vaults of the cross vaults in four types, distinguishing for each of them according to the plan geometry (regular plan and rectangular plan). The first cross vault surfaces described are those with cylindrical caps, followed by surface of the cross vaults with cylindroid caps; surface of the cross vaults with spherical caps generated by a circular arc of variable radius; surface of the cross vaults with spherical caps generated by a circular arc of constant radius.

In the ribbed cross vaults of our case study, the geometry derived from the point clouds suggests a shape very close to that exemplified by the surface of the cross vaults with cylindroid caps proposed by Curioni. Describing the surface of the cross vaults with cylindroid panels, Curioni assumes that the perimeter arches are circular and the diagonal ones are elliptical arches. He defines the intrados surface as generated by a straight line that moves through a point of the circumference and a point of the ellipse remaining parallel to the vertical plane determined by the ridge line (Fig. 4). In our case study, the geometry of the diagonal arches of the vaults is close to that of the circle since the measured value of the difference between the two axle shafts is of the order of a few centimeters. For the application of the geometric model proposed by Curioni, we therefore decided to consider the geometry of the circle assuming it as an ellipse with zero eccentricity. As shown in the following paragraph, the geometry of Curioni's cylindroid web shell is well translated by the algorithmic modeling adopted.

3 Algorithmic Modelling for Complex Vault Systems

In the algorithmic modeling, the theoretical foundations and the mathematical descriptions enunciated in the manuals and in the architectural treatises can constitute a useful reference and an effective starting point. The same procedures proposed by Curioni for computing the surfaces of the cross vaults can be conducted algorithmic modeling by analyzing the individual caps as separate elements, and then assembling them together to make up the complete vault.

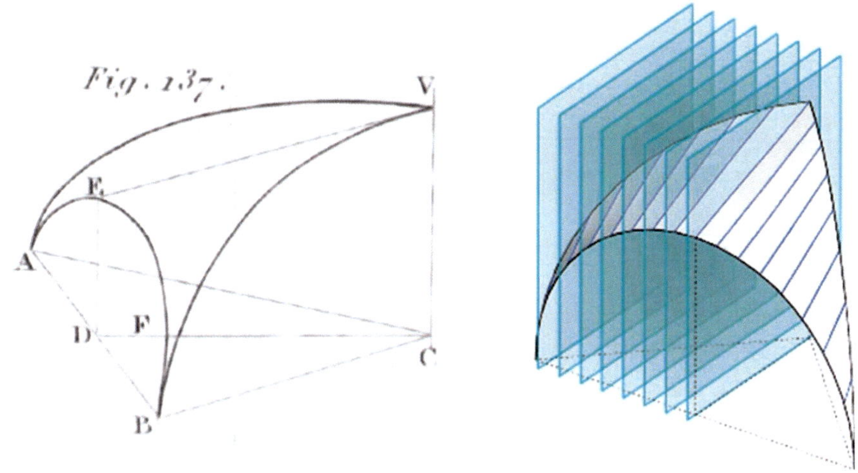

Fig. 4 The illustration of the cylindroid surface proposed in Giovanni Curioni's "L'arte di fabbricare" (on the left) and the graphing of the genesis transposed in the construction of the NURBS (on the right)

The modelling of complex historical elements, like these caps, in order to implement them in HBIM processes presents us with some significant problems (Bagnolo et al. 2019b). First and foremost is the lack of prebuilt parametric objects/families, which therefore requires the use of ad hoc local modelling; the local models obtained have little or no margin for modification, thus rendering the models unusable for contexts and elements different from the initial ones. In the case in which we want to create reusable models through the parametrization of the elements, we often must face limitations given by the little flexibility offered by the native modelling tools present in BIM environments. The solution to this is the use of external tools, more complex and flexible, which allow you to still obtain local objects and therefore require a continuous passage between different work environment.

An alternative way to the normal Scan-to-BIM workflow is given by the insertion of algorithmic modelling tools within this process. In fact, they allow to develop a sequence of instructions, even if not limited to the modelling field alone, which in fact makes the process itself parametric, where this is difficult to obtain on the single element; specifically, it was decided to use Autodesk's Dynamo software, an extension included within Revit (Argiolas et al. 2019). This allows you to code the steps that lead to the generation of the model by means of command blocks which can be linked together, and which can be modified and managed by entering the input parameters. The developed algorithms are not strictly linked to the single object to be modelled, as much as to the type of element, to its geometric genesis (Figs. 5 and 6).

This approach reflects the attempt to codify the geometric construction rules illustrated in the historical treatises; in fact it is not so wrong to see the geometric patterns that can be found in historical treatises such as sequences of rules and steps for the

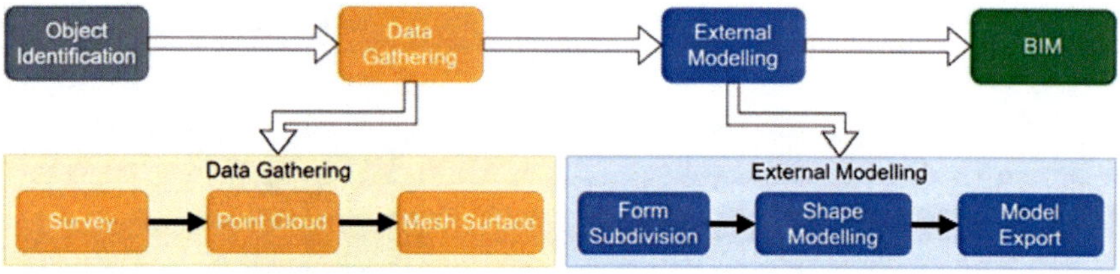

Fig. 5 Standard scan-to-BIM workflow

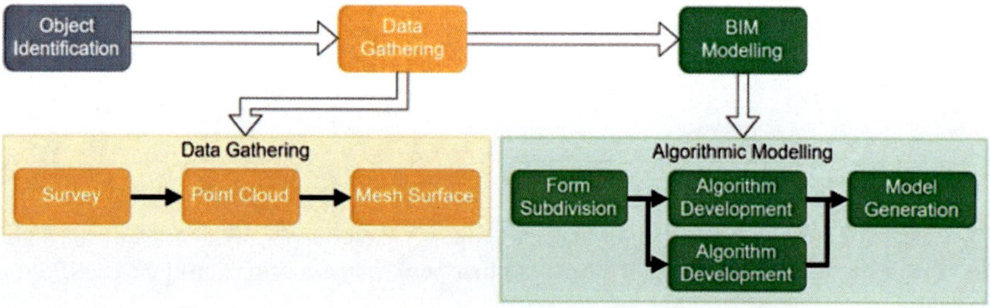

Fig. 6 Scan-to-BIM workflow with algorithmic modelling implementation

generation of "ideal" models of architectural objects in order to represent real ones. This parallelism is in fact one of the strengths of the application of algorithmic modelling, as it combines the advantage of reusability and flexibility with the possibility of storing information on the genesis of the elements within the algorithm. The process itself therefore becomes a container of information, like the object created, according to the dictates of the BIM philosophy.

The work presented is based on the desire to explore the possibility of implementing complex elements in Scan-to-BIM flows, coding their modelling according to the rules drawn from historical treatises; rules that are analysed and reinterpreted so that they can adapt to the new digital tools of graphic programming. As already mentioned, in the modelling of cross vaults, Curioni's contribution to the classification of the genesis and the geometric characteristics of the surfaces from which they sometimes take shape is invaluable. Curioni in fact compiles a list of case studies based on the characteristics of the space to be covered and the surfaces that are applied to it. In particular, the treatise offers a sequence of development of the geometries starting from the curves, elements at the base of the development of the surfaces, and goes to generation of simple and composed vaults. This approach is similar to that necessary for the development of a computer algorithm in which, starting from the basic elements such as points, lines and arcs, these become gradually input for the blocks dedicated to the modelling of the increasingly complex elements (Fig. 7).

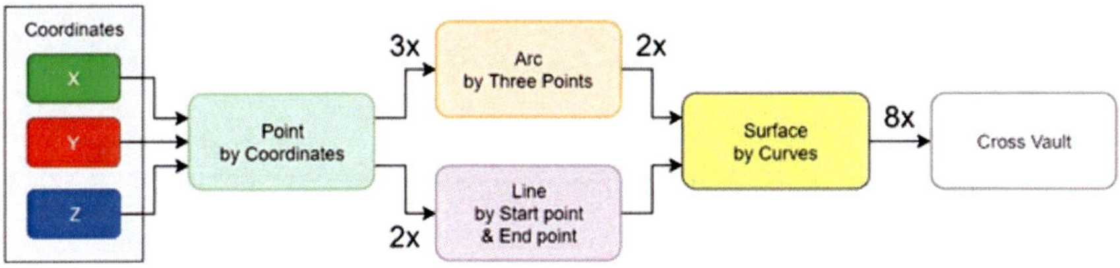

Fig. 7 The algorithm for the modelling a cross vault starting from point

4 Background

Since the coining of the term HBIM (Murphy et al. 2009) one of the main problems of applying the BIM methodology to historical heritage is to implement a parametric modelling of historical elements. Often this modelling technique can be complicate due to the causes of the native instruments of the BIM environments (Oreni et al. 2014); this leads to having to use modelling tools external to BIM for more complex objects (Barazzetti et al. 2016). In cases where parametric modelling is possible, the results are often too tied to the geographical context, preventing the use of families in other areas (Baik 2017).

The possibility of using parametric modelling systems for the parametric generation of complex elements is only recently being explored (Capone and Lanzara 2019). Algorithmic (or procedural) modelling tools such as GrassHopper for Rhino guarantee to parameterize the process of generating objects, before their import in the BIM environment, partially solving the problems related to complex geometries, but leaving unchanged the need to resort to instruments external to BIM. Another possible way is to employ internal tools such as Revit Dynamo, which guarantees to maintain the entire modelling process within the BIM environment.

5 Case Study: Rib Vault System in the Church of Nostra Signora Della Speranza in Cagliari

The vaulted system covering the nave of the church of Nostra Signora della Speranza, setting on a rectangular plan, has a longitudinal extension of about m and a width of 4. m; the central cross vault turns out to have a double dimension compared to the lateral half vaults. All vaults of the system have pointed perimeter and diagonal arcs with central keystones at a height raised with respect to the keystones of the perimeter arcs. The ridge lines joining the keystones of the arches with the main keystone of the vaults are straight, a typical feature of the surfaces. Curioni calls this surfaces cylindrical but the vaults under examination present the peculiarity of having both non-straight profiles that can be traced back to arcs of circumference and not elliptical, differentiating them from the constructions described by Curioni (Wendland

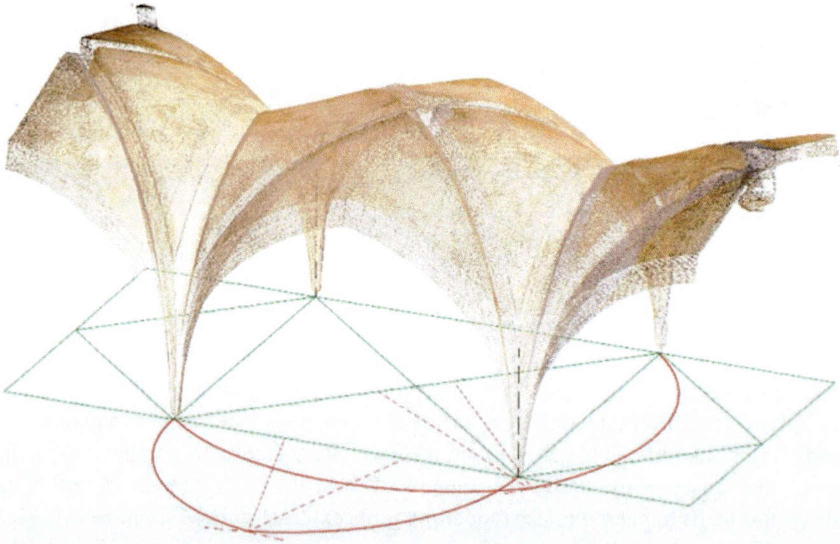

Fig. 8 Point cloud with RGB coordinates on the geometric scheme of the vaults

2005). This "pseudo-cylindrical" configuration with a rectilinear ridge is due to the quadripartite squares on a square plan deriving from the transformation of the cruise (Capone 2014) (Fig. 8).

Resuming the logic followed by the writer, according to which it is possible to study the vaults as a union of individual caps, it was decided to base the modelling in the same way, proceeding then to model the individual portions of surfaces and then assemble them together. The generation of each surface is done by defining the three profiles that delimit it, in turn defined by three points, the minimum number necessary to uniquely identify the arcs of circumference. According to this methodology for the definition of each cap six points are required, with a total of 15 points for the modelling of half cross vault and 25 points for modelling of the whole central cross vault.

It is therefore possible to define a lattice scheme defined by the straight lines passing through the extremes and midpoints of the described arcs, whose intersections identify the positions of the points whose elevations will be given as input parameters of the algorithm. Once the modelling method has been defined, it was decided not to set in this initial stage a procedure for the automatic matching of individual vaults, an operation that is currently set manually.

The coordinates of the vertices of the cruises projected on the spring plane are the input parameters of the algorithm, used for the automatic generation of the planimetric schemes on which the previously described reference points are identified. The values of the heights of each point are passed to the algorithm as further parameters allowing the tracing of the arcs delimiting the caps and consequently the creation of the surfaces themselves. The model thus obtained consists of 16 NURBS surfaces, each representing a half-cap, which can be imported directly into Revit as a "import object" for direct use or for inclusion in a family (Figs. 9 and 10). In Revit it is possible to make a first comparison between the NURBS model and the point cloud; this can be done through the main sections such as the one passing through the perimeter arches, the main keystone and the diagonals (Figs. 11, 12 and 13).

Next step is the validation of the generated model to verify the degree of correspondence with the point cloud resulting from the survey using the laser scanner; the validation takes place using the CloudCompare software, which allows the computation of the differences between surfaces and point clouds. For the model to be set up in CloudCompare, a conversion to a mesh surface is required; this operation is

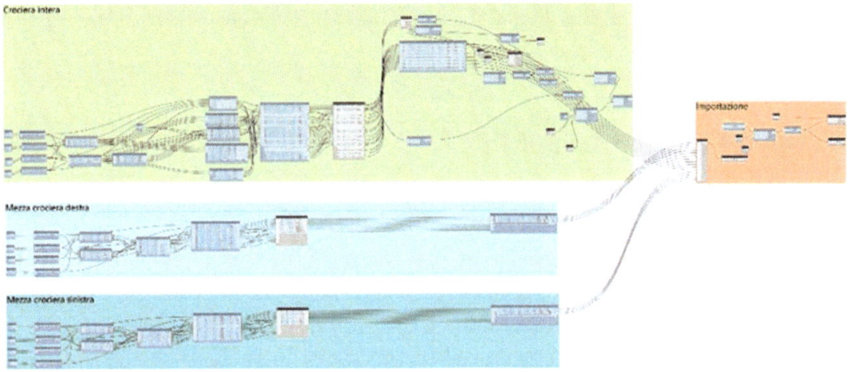

Fig. 9 Dynamo algorithm for the modelling of the vaulted system

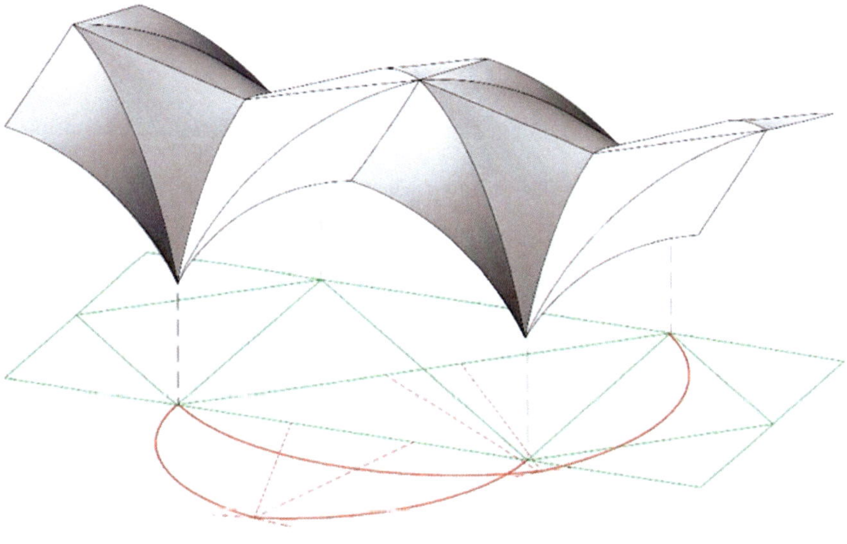

Fig. 10 Algorithmic model on the geometric scheme of the vaults

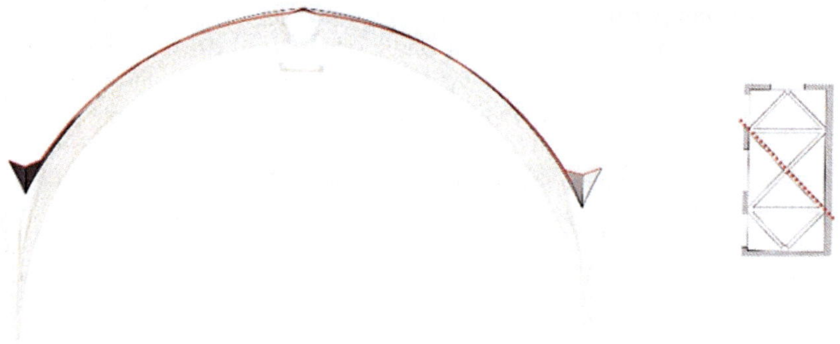

Fig. 11 Comparison between the diagonal section of the NURBS model (in red) and that deriving from the point cloud

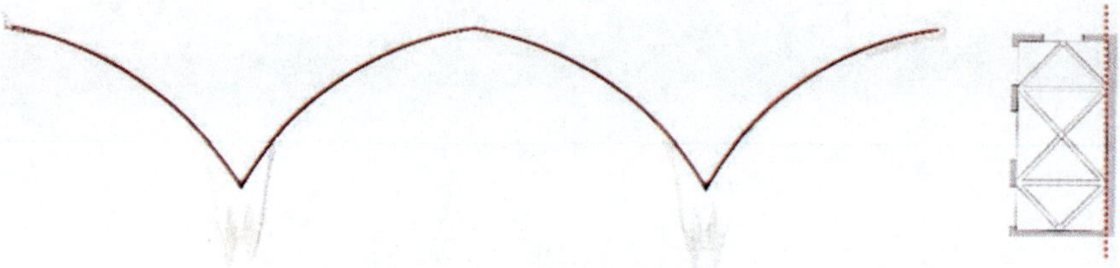

Fig. 12 Comparison between the section of the NURBS model (in red) and that of the point cloud along the formeret (wall rib)

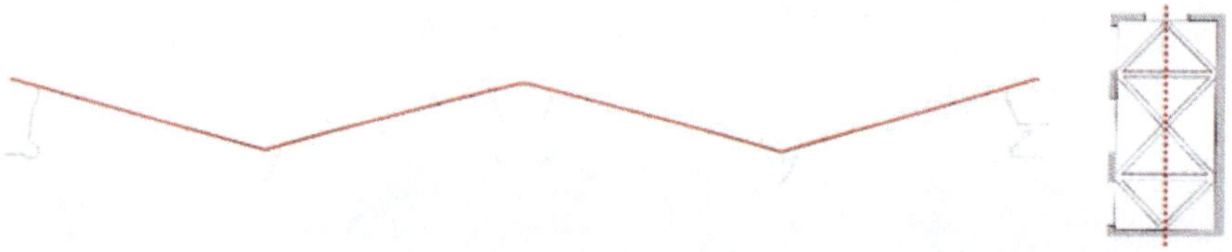

Fig. 13 Comparison between the longitudinal section of the NURBS model (in red) and that of the point cloud along the ridge rib

easily performed in Dynamo thanks to NURBS surface approximation tools using mesh surfaces, which are then exported for external use.

Point cloud and mesh surface are imported in CloudCompare and their alignment is carried out, then the deviations are quantified, in order to understand the reliability of the model generated algorithmically and to locate any critical issues. From the results provided by the comparison it is possible to note that the surfaces have a correspondence such as to limit the discrepancies to values less than 2 cm, while higher deviations are in the areas corresponding to the ribs and the springs of the vaults (Fig. 14). This result reflects what was expected since, having decided to

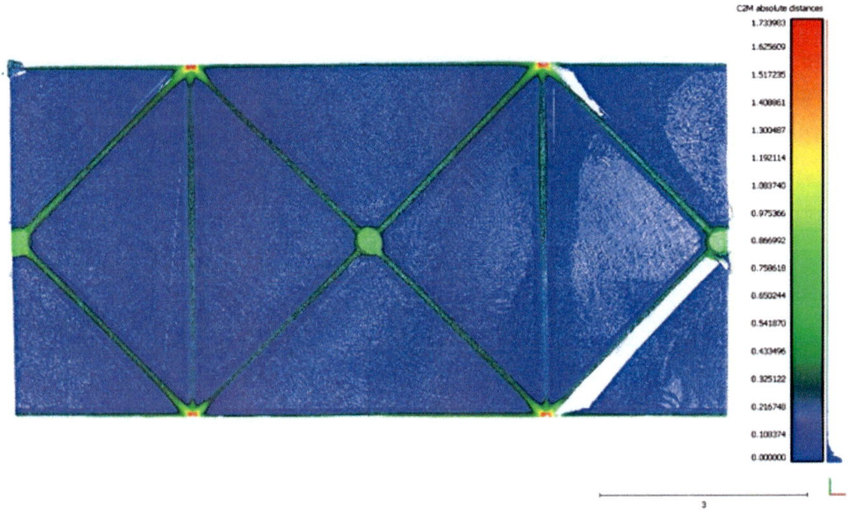

Fig. 14 Comparison between the algorithmic model and the Point Cloud from laser scanner

not consider the ribs in this research stage, these were excluded in the modelling operations. This is also the cause of the discrepancies in the springs, hidden in the point cloud by the ribs.

6 Conclusions

At this point of the research, we have come to develop an algorithm by which it is possible to generate the surface of each single panel making up of quadripartite ogival vaults. We can verify the quality of the results by comparing the NURBS and the TLS data through the Cloudcompare software. The comparison between the two models shows the effectiveness of the proposed algorithm, obtaining a value of the maximum deviation which does not exceed 2 cm.

Algorithmic modelling proves to be a valid improvement in standard Scan-to-BIM workflows, offering a possible solution to issues still open like low flexibility of native 3D modelling tools of BIM environments, single-use modelling processes and hard to modify methods. The parametrisation of the generation process offers new ways to interact with the process itself, limiting inputs constrains and making the implementation of new features easier.

The developed algorithm still has a huge margin of improvement since some of the actual functions can be automated and optimized. The assembly of halved and full vaults in sequences are set manually in this stage, but algorithmic modelling allows to think a string driven process in which the users can create specify the type of elements, halved or full, or the orientation; then the algorithm will generate and place the various parts in the right position.

Another automation should be made for profiles recognition, so that the arcs delimiting surfaces can be defined using more points overcoming the limit of circular arcs, allowing to use other conic curves. Some aspects have to be investigated in further stages, in first place the implementation of the ribs in the modelling process.

Writing the algorithm of the cross vaults of this case study was particularly effective as the detected geometry corresponds to a graphic primitive whose procedural modelling is easily manageable. The research obviously requires the analysis of further case studies comparable with those discussed here, as well as the expansion of research on historical treatises in order to implement other reference models.

Acknowledgements The 3D laser scanning of the vault was carried out at LabMAST (Laboratory for historical and traditional materials and architectures), University of Cagliari, Department of Civil-Environmental Engineering and Architecture (DICAAR). LIDAR point cloud product by Sergio Demontis and Valentina Pintus.

The editorial responsibility of the paragraphs is recognized to: V. Bagnolo for paragraphs 1, 2 and 6, R. Argiolas for paragraph 3, 4 and 5.

References

Argiolas R, Cazzani A, Reccia E, Bagnolo V (2019) From LIDAR data towards HBim for structural evaluation. Int Arch Photogram Remote Sens Spatial Inf Sci XLII-2/W15:125–132. https://doi.org/10.5194/isprs-archives-XLII-2-W15-125-2019

Bagnolo V, Argiolas R (2019a) 3D procedural modeling of complex vaulted systems. Geometric rules vs SfM based modeling. In: 2019 IMEKO TC-4 international conference on metrology for archaeology and cultural heritage (MetroArchaeo 2019), pp 211–216

Bagnolo V, Argiolas R, Cuccu A (2019b) HBim for archaeological sites: from SfM based survey to algorithmic modeling. Int Arch Photogram Remote Sens Spatial Inf Sci XLII-2/W9:57–63. https://doi.org/10.5194/isprs-archives-XLII-2-W9-57-2019

Baik A (2017) From point cloud to Jeddah heritage BIM nasif historical house—case study. Digital Appl Archaeol Cult Heritage 2017(4):1–18

Barazzetti L et al (2016) BIM from laser scans… not just for buildings: NURBS-based parametric modeling of a medieval bridge. In: XXIII ISPRS congress, commission V, pp 51–56

Capone M (2014) La discretizzazione della forma. Genesi e trasformazione: la geometria segreta dei reticoli spaziali delle volte gotiche. Disegnare idee immagini, anno XXV 49:36–48

Capone M, Lanzara E (2019) Scan-to-BIM vs 3D ideal model HBim: parametric tools to study domes geometry. Int Arch Photogramm Remote Sens Spatial Inf Sci XLII-2/W9:219–226. https://doi.org/10.5194/isprs-archives-XLII-2-W9-219-2019

Curioni G (1869) Geometria pratica applicata all'arte del costruttore. Negro, Torino

Murphy M, Mcgovern E, Pavia S (2009) Historic building information modelling (HBIM). Struct Surv

Oreni D, Brumana R, Banfi F, Bertola L, Barazzetti L, Cuca B, Previtali M, Roncoroni F (2014) Beyond crude 3D models: from point clouds to historical building information modeling via NURBS. Euro-mediterranean conference. Springer, Cham, pp 166–175

Runne H, Niemeier W, Kern F (2001) Application of laser scanners to determine the geometry of buildings. In: Grün A, Kahmen H (eds) Optical 3D measurement techniques. Vienna University of Technology, Vienna, pp 41–48

Tang P, Huber D, Akinci B, Lipman R, Lytle A (2010) Automatic recon- struction of as-built building information models from laser-scanned point clouds: a review of related techniques. Autom Constr 19(7):829–843. https://doi.org/10.1016/j.autcon.2010.06.007

Thomson C, Boehm J (2015). Automatic geometry generation from point clouds. Remote Sens 7:11753–11775. ISSN: 2072-4292

Vitali M, Natta F (2019) Digital survey and 3D geometric interpretation of complex vaulted systems (Palazzo Valperga Galleani di Barbaresco in Turin). In: 2019 IME-KO TC-4 international conference on metrology for archaeology and cultural heritage (MetroArchaeo 2019) proceedings. Florence, Italy, pp 205–210

Wendland D (2005) Some considerations on the shape of the caps of vaults. In: Modena C, Lourenco PB, Roca P (eds) Structural analysis of historical constructions. Taylor & Francis, London, pp 111–120. ISBN: 04 15363799

Mixed Reality Experiences for the Historical Storytelling of Cultural Heritage

Stefano Brusaporci, Fabio Graziosi, Fabio Franchi, Pamela Maiezza, and Alessandra Tata

Abstract The aim of the research is the realization of Mixed Reality experiences to narrate the history of architectural heritage. This work is related to the INCIPICT Project and to 5G experimentations conducted by L'Aquila University for the supporting of scientific research and productive activities through the realization of a smart city. In particular, the work roots on the 3D reconstruction of the no-more existing Baroque configuration of the Basilica of Collemaggio in L'Aquila (IT), made according to the analysis of archival documents and the surveying of the current building, with the integration of laser scanning and photogrammetry by drone pictures. The Baroque system was demolished in the Seventies for a stylistic restoration. ICT allows the ubiquitous visualization of real time renderings of the old historical phase on people's personal devices, coherently superimposed to the surrounded reality framed by the users. The 3D model's visualizations have an educational purpose, but also configure as visual-computing tool for historical-critical study and enhancement of architectural heritage.

S. Brusaporci (✉) · P. Maiezza · A. Tata
Department of Civil Construction-Architectural and Environmental Engineering, University of L'Aquila, L'Aquila, Italy
e-mail: stefano.brusaporci@univaq.it

P. Maiezza
e-mail: pamela.maiezza@univaq.it

A. Tata
e-mail: alessandra.tata@graduate.univaq.it

F. Graziosi · F. Franchi
Department of Information Engineering, Computer Science and Mathematics, University of L'Aquila, L'Aquila, Italy
e-mail: fabio.graziosi@univaq.it

F. Franchi
e-mail: fabio.franchi@univaq.it

C. Bolognesi and D. Villa (eds.), *From Building Information Modelling to Mixed Reality*, Springer Tracts in Civil Engineering,
https://doi.org/10.1007/978-3-030-49278-6_3

Keywords Mixed reality · Architectural heritage · Architectural survey · 3D modeling

1 Introduction: The INCIPICT Project and 5G Experimentation for Architectural Heritage Enhancement in L'Aquila City

The study roots in the INCIPICT Project (Innovating City Planning through Information and Communications Technology) of L'Aquila University (http://incipict.univaq.it), financed by the CIPE (Interministerial Committee for Economic Planning) of the Italian Government, after the earthquake of 2009 that has severely hit the city of L'Aquila (Italy), and caused several collapses of buildings and more than 300 victims.

The INCIPICT Project integrates with the 5G experimentation, in progress in L'Aquila city in collaboration with Wind and ZTE. INCIPICT + 5G studies aim to support the scientific research and productive activities in L'Aquila through the use of ICTs. Therefore, one of the main outcomes is the realization of a smart city, where the technological opportunities merge and match with cultural issues. In this way, the L'Aquila city itself configures as a living laboratory: the city is intended as a complex organism of tangible and intangible aspects, including the historical buildings, the cultural and social context, the economic system, the network of infrastructures and services (Brusaporci et al. 2018a).

With specific reference to the theme of cultural heritage enhancement, the research unit is working on the use of augmented reality to narrate the history of architectural heritage and places, through the visualization of no more existing past configurations (Brusaporci et al. 2017, 2018b, 2019).

We present an experimentation on the virtual reconstruction of the Baroque configuration of Collemaggio Basilica in L'Aquila. At the beginning of the Seventies, a stylistic restoration destroyed the Baroque apparatus to bring the church back to a presumed medieval guise, according to preexistences discovered into the masonries and under the stuccoes. Moving from the digital surveying, through the study of historical graphical and photographical documents, a 3D photorealistic model of the no more existing baroque church has been realized, aiming to digital in site visualization for the storytelling of a particular phase of the history of the Basilica (Cavazza and Donikian 2007; Clini et al. 2017) (Fig. 1).

2 The History of Collemaggio Basilica

The Collemaggio Basilica in L'Aquila was founded in 1287, probably completed in its first configuration in 1294, when Pietro del Morrone was here crowned Pope with the name Celestino V.

Fig. 1 Basilica of Collemaggio, L'Aquila: the facade

By the end of the fourteenth century, many transformations—also consequently the earthquakes of 1316 and 1349—have led to the current settlement with three-naves, transept in line with the fabric, flat terminated apses. The current rectangular facade can be traced back to the fifteenth century (Gavini 1980; Giardini et al. 2006).

As a result of the damages caused by an earthquake in the mid-fifteenth century, important works began; but it is above all by the 17th century that the baroque interior reconfiguration takes place, with an overall re-configuration completed in 1669 for the main hall (Moretti 1971; Antonini 1999).

On the baroque configuration and decoration, some authors underline the local influences (Colapietra II 1158), other Central European ones (Moretti 1972a), but more likely the most important influences are correlated with southern cultural flows related to other important Celestinian centers (Antonini, 201).

The earthquake of 1703 seriously damaged the apses and the transept, and to the subsequent interventions of the XVIII century are attributable the decorations and the altars of the apses and of the transept, and the vaults on the side aisles (Antonini 1999; Colapietra 1978) (Figs. 2 and 3).

Between 1970 and 1972 the Superintendent Mario Moretti realized a stylistic restoration, aimed to the "re-discovering" of the medieval building, hypothetically preserved under the surface of the baroque "superfetations". The approach is based on a prejudice of stylistic value towards artistic and cultural expressions chronologically more recent than the ancient one, and on the mistaken belief that baroque interventions were simple overlaps that had not affected the medieval "support". From this cultural line, numerous stylistic restoration have been realized in Abruzzo

Fig. 2 View of the main nave of the Basilica of Collemaggio

Fig. 3 The survey of the basilica: plant and longitudinal section of the point cloud

(Torlontano 2010). Practically, the Moretti intervention of restoration is operationally based on the "research" of pre-existing medieval elements through demolitions, with the removal of "modern" walls and baroque elements (plasters, stucco, altars, cornices, false ceilings). Once he discovered partial surviving medieval elements, an overall redesign of the configuration of the building followed. The restoration of Collemaggio had a wide media impact, nourishing a bitter discussion, with critical interventions also of well-known figures as Bruno Zevi (Moretti 1972b).

Only the eighteenth-century configuration of the apses and the transept of the Baroque apparatus survives to the restoration of Moretti.

Finally, the 2009 earthquake once again caused serious damage to the basilica, particularly in the area of the transept and apses, even with major collapses. This was followed by a critical conservative restoration intervention; the Basilica was re-opened in 2017.

3 The Virtual Reconstruction of the Baroque Main Nave: Surveying, Methodology, and Outcomes

The research moves from the surveying of the current configuration of the church, intended as the only architectural document really experienceable, and from the study of the old documents, in particular the graphical surveying of the church before the stylistic restoration (plant, longitudinal and cross sections), and the old photos taken before and during the works. The old drawings present a low level of detail, and the baroque decorations are identically repeated; many particulars are different from the photographic ones. Moreover, the old photos have a low resolution; many times is difficult to define the exact element framed by the camera—also because they are non-more existing. Aiming to define the baroque particulars, inverse prospective restitutions (made in a graphical way or using digital photogrammetry inverse camera procedures, such as in forensic applications) are of little or no use (Figs. 4 and 5).

A specific issue is related to the definition of the colours of the baroque plaster, stuccos, decorations, also of the wooden false ceiling; only a coloured picture exists, probably recoloured, but at the same time as when it was taken, probably in the Sixties: therefore, it is a very important document. All the historical documents are saved in the historical archives of Abruzzo Superintendence, and many of these are published (Moretti 1972a).

The architectural surveying of the Basilica was realized with the integration of Leica BLK360 laser scanner and digital photogrammetry applied to pictures taken by a DJI Phantom 4 drone. UAV technology have been necessary to realize the point cloud of the external parts of the building that cannot be measured by the terrestrial scanner, in particular the roofs, the façade and the related architectural particulars.

Using the drone, a dataset of 159 images of the exterior, 86 images of the so-called Holy Entrance, 219 of the main façade, and 25 of the main rose window have been registered. According to Structure from Motion technology, images have

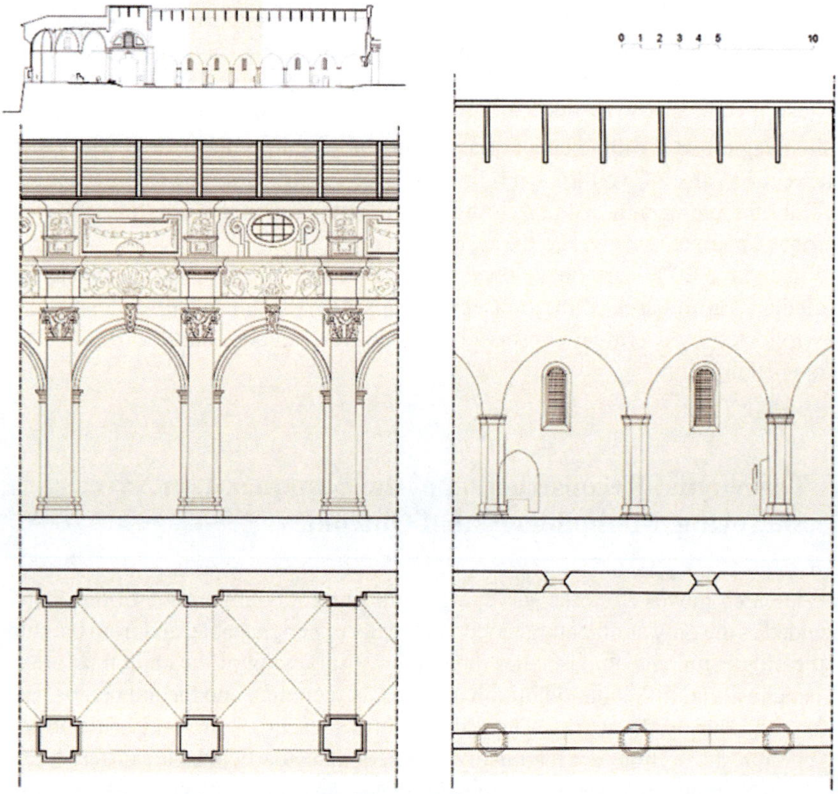

Fig. 4 Comparison between the re-drawing of the baroque configuration and the survey of the current state

been elaborated with the Agisoft PhotoScan software, realizing a point cloud and a textured mesh. The laser scanning campaign was realized with 37 station points into the church and 13 outside. The instrumental resolution between two points at 10 m is of 20 mm. The 50 scans have been recorded with Autodesk ReCap software. The terrestrial laser scanner point cloud and the aerial UAV point cloud have been registered with Cloud Compare software. The whole cloud, imported in Autodesk Autocad, has been used to draw current bi-dimensional restitution of the Basilica, such as plants, sections, and elevations (Bianchini 2014).

According to this archival information, the baroque configuration has been critically re-designed in relation to the existing building, first in bi-dimensional graphics, then in the three-dimensional space, with the use of point clouds and 3D models. The three-dimensional model of the Baroque church has been realized with Rhinoceros software (Fig. 6).

Fig. 5 Digital reconstruction of the baroque configuration: view of the main nave

This spatial work favors a "visual computing" processes of understanding: it is a critical and iterative process, where the modeler, observing and re-drawing the documents, creates his own ideal concept; then he elaborates his idea and refines it though the modeling; consequently, he compares it back to the documents, and the process restarts, evolving toward the final critical interpretation. Therefore, the 3D restitutive model configures as a new critical contribution of knowledge (Brusaporci 2015; Apollonio 2012; Maiezza 2019) (Fig. 7).

This 3D modeling workflow is based on an interpretative reflection and reconstruction, without automatic processes: the baroque elements have been shaped in analogy and proportion to the ones represented in the by old documents: it is an artisan approach and attention, even if conducted in the digital space and with advanced modeling tool. This works roots on a digital molding skill—similar to that of the artisans who worked in the ateliers and in the pre-modern construction sites—that require the knowledge of the past design rules and technologies by the contemporary digital "handicrafts" worker (Guidi et al. 2014).

The aim of the research is not the apologia of the disappeared baroque configuration, but a study based on an historical-critical methodology, the understanding of historical and aesthetical values rises from the reconstruction of the history.

Therefore, the study aims to reconstruct and narrate the physical transformations, that is a phase of the history of the Basilica of Collemaggio, which derives the current configuration of the church, important both from a historical and cultural point of view.

Fig. 6 Rendered view of the baroque configuration

The model also favors the description and valorization of the current church: it is an educational purpose and it presupposes an active participation of users in developing a personal critical reflection, based on the new cognitive outcome offered by the visualization of the 3D model in relation to the physical church (Luigini 2017, 2019).

Fig. 7 Detail view of the baroque apparatus reconstructed in the 3D model

4 Mixed Reality Application

Mobile augmented reality falls within the boundaries of augmented reality (AR) re-search. The initial description of what AR is can comes from Sutherland's 1965 essay (Sutherland 1965) A more technical definition for AR is provided by Azuma (1997). This definition is a basis for discussing AR in precise terms and it defines the following three properties for AR:

It should combine the real and the virtual.
The augmentations should be interactive in real time.
They should be registered in three dimensions.

In contrast to virtual reality (VR), AR does not replace the real world with a simulated world (Steuer 1992). Rather, it seeks to combine the real with the virtual, with the virtual (augmentations) being interactive in real time and in three dimensions.

With the recent proliferation of mobile devices incorporating more processing power and features, including GPS, accelerometers, gyroscopes, and advanced computer vision algorithms, AR has become both feasible and affordable, leading to its widespread adoption.

The fundamental difference between Augmented and Virtual Reality consists in the concept of simulation used. Virtual reality leads us, through a more or less immersive system, to think of living a certain reality deceiving our senses; this reality is completely computer-generated. So, the VR stands between us and the real world, breaking at several levels sensory communication with it and replacing it entirely with a fictional environment.

Augmented reality, on the contrary, takes advantage of the real world as the basis on which to add information layers. "Augmented" refers to the feeling of "enhancement" of perception, understood as an expansion of the information that we would normally perceive using our senses. This technology is achieved in each case creating virtual content which, exactly as in virtual reality, aims at providing visual, auditory and even olfactory and tactile data, integrating them in the commonly perceived real space.

The 3D contents related to Collemaggio Basilica in L'Aquila are used to extend the tourist experience by exploiting the potential AR solutions. Since the main purpose for AR is to enhance the reality with virtual content, it is important to make sure that virtual objects are correctly registered to the real scene. This can help users view the virtual content as part of the real world. Correct registration can be obtained by estimating the pose of the camera (for video see-through) or user's viewpoint (for optical see-through). The registration process usually consists of two parts. In the first part, fiducial markers or feature points are detected, using marker-based methods or marker-less methods (Huang et al. 2013) Then the second part estimates the pose and maps 3D virtual objects through proper projective geometry. The maker-less method is the one chosen for the experimentation. It helps the AR system detects the scene in a more natural way. In the feature extraction step, the goal is to find areas of interest in the input image that can be served as unique and reliable markers. There are lots of feature detection and extraction algorithms based on different single or combination of features, such as edges, corners, blobs, ridges (Deriche and Giraudon 1993).

The implementation of an AR system required the development of a computer vision system that is able to recognize objects, places or details allowing the user to interact with them or gather information. Concerning the development of the solution for the experimentation, the focus is toward mobile devices, such as tablets and smartphones, in order for the users to exploit their own devices. Nevertheless, AR glasses have also taken into account (Fig. 8).

5 Conclusion

This interdisciplinary study rises from the collaboration of computer science, ICT, and cultural heritage scholars. The outcome is the realization of a mixed reality cultural experience. From a methodological point of view, the work moves from the concept that reality is the result of processes of transformation and modification during time, and that the 'narration' of these processes favors the understanding of the 'reasons' and 'values' of buildings and places. On the History for the Architecture, Spagnesi (1984, p. 7) wrote: "If doing the 'history' is always equivalent to knowing, the History of Architecture can only be the knowledge of the physical space built by man, that is to say the current reality. Therefore, wishing to pose the problem of

Fig. 8 Mixed reality with the overlapping of the old Baroque configuration on the current view

'knowing' today the 'reality', we only have to analyze the occurrence of the essential reasons that produced it in a temporal succession, as they occur, and with respect to which homogeneous time periods in the history of human communities".

At the same time, the digital multimedia 'artifact' have the status of cultural expression of human doing (UNESCO 2003), and together with the reality to which they are referred, it is a new kind of cultural heritage. In fact, digital heritage from real contents—in particular in mixed reality applications—is not something opposed to the "real", but it is a further manifestation related to the tangible, which can play an important role in processes of interpretation, communication, conservation and enhancement of cultural heritage (Brusaporci and Maiezza 2018; Trizio et al. 2019; Russo et al. 2019; Luigini et al. 2019).

In particular, the 5G technology allows managing a large number of mobile devices, connected with a low latency, i.e. with an extremely reliable and fast response speed, allowing a ubiquitous massive flow of data. In the cultural heritage field, this favors virtual and augmented reality applications in real time. In this way, it is possible to see on the screen of the smartphone—or other kind of devices, renderings of complex informative models in such a short time to allow the coherent superimposition of these images to those of the buildings or places observed and framed by the user (Lackey and Shumaker 2014; Dragoni et al. 2018). Therefore, people are free to move in the space that surrounds them, synchronically observing both the reality and the information that is visually superimposed. In this way, we have an effective advanced Internet of Things system, but realized with an overall respect of the historical "matter" of the works of art, where the interaction between technological applications and heritage physicality occurs only in the dimension of the digital image.

Acknowledgements The research has received funding from the Italian Government under Cipe resolution n. 135 (Dec. 21, 2012), project INnovating City Planning through Information and Communication Technologies (INCIPICT).

References

Antonini O (1999) Architettura religiosa aquilana, vol 1. Edizioni del Gallo Cedrone, L'Aquila

Apollonio FI (2012) Architettura in 3D. Modelli digitali per i sistemi cognitivi, Mondadori Bruno, Milano

Azuma RT (1997) A survey of augmented reality. Presence Teleoperators Virtual Environ 6(4):355–385

Bianchini C (2014) Survey 2.0: new technologies, new equipment, new surveyors? In: Giandebiaggi P, Vernizzi C (eds) Italian survey & international experience. Gangemi Editore, Roma, pp 763–768

Brusaporci S (2015) On visual computing for architectural heritage. In: Brusaporci S (ed) Handbook of research on emerging digital tools for architectural surveying, modeling, and representation, vol 1. IGI Global, Hershey, PA, pp 94–123

Brusaporci S, Ruggieri G, Sicuranza F, Maiezza P (2017) Augmented reality for historical storytelling. The INCIPICT project for the reconstruction of tangible and intangible image of L'Aquila historical centre. Proceedings 1(9):1083

Brusaporci S, Maiezza P (2018) Tra Storia e Memoria. Tecnologie avanzate per la (ri) definizione partecipativa del significato dei luoghi nella città storica. In: Luigini A, Panciroli C (eds) Ambienti digitali per l'educazione all'arte e al patrimonio. Franco Angeli SRL, Milano, pp 51–63

Brusaporci S, Maiezza P, Tata A (2018a) For a cultural-based smart city. In: Salerno R (ed) Rappresentazione materiale/immateriale—drawing as (in) tangible. Gangemi Editore, Roma, pp 73–80

Brusaporci S, Graziosi F, Franchi F, Maiezza P, Vernacotola F (2018b) La memoria dell'effimero e la contingenza del precario. In: Capano F, Pascariello MI, Visone M (eds) La Città Altra. Storia e immagine della diversità urbana. FedOA—Federico II University Press, Napoli, pp 1185–1192

Brusaporci S, Graziosi F, Franchi F, Maiezza P (2019) Remediating the historical city. Ubiquitous augmented reality for cultural heritage enhancement. In: Luigini A (ed) Proceedings of the 1st international and interdisciplinary conference on digital environments for education, arts and heritage. EARTH 2018. Springer, Cham, pp 305–313

Cavazza M, Donikian S (eds) (2007) Virtual storytelling using virtual reality technologies for storytelling. Springer, Cham

Clini P, Quattrini R, Frontoni E, Pierdicca R, Nespeca R (2017) Real/not real: pseudo-holography and augmented reality applications for cultural heritage. In: Ippolito A, Cigola M (eds) handbook of research on emerging technologies for digital preservation and information modeling. IGI Global, Hershey, PA, pp 201–227

Colapietra R (1978) L'Aquila dell'Antinori: strutture sociali ed urbane della citta nel Sei e Settecento. In: Colapietra R (ed) Antinoriana Studi per il bicentenario della morte di Antonio Ludovico Antinori. Deputazione Abruzzese di Storia Patria, L'Aquila

Deriche R, Giraudon G (1993) A computational approach for corner and vertex detection. Int J Comput Vision 10(2):101–124

Dragoni AF, Quattrini R, Sernani P, Ruggeri L (2018) Real scale augmented reality. A novel paradigm for archaeological heritage fruition. In: Luigini A (ed) proceedings of the 1st international and interdisciplinary conference on digital environments for education, arts and heritage. EARTH 2018. Advances in intelligent systems and computing, vol 919. Springer, Cham, pp 659–670

Gavini IC (1980) Storia dell'Architettura in Abruzzo, vol 2. 1st edn. Costantini Editore, Pescara (1927–28)

Giardini L, Pezzuti M, Redi F (2006) Celestino V e la sua basilica. Silvana Editoriale, Milano

Guidi G, Russo M, Angheleddu D (2014) 3D survey and virtual reconstruction of archaeological sites. Digit Appl Archaeol Cult Herit 1(2014):55–69

Huang Z, Hui P, Peylo C, Chatzopoulos D (2013) Mobile augmented reality survey: a bottom-up approach. Preprint at arXiv:1309.4413

Lackey S, Shumaker R (eds) (2014) Virtual, augmented and mixed reality: applications of virtual and augmented reality. Springer, Cham

Luigini A (ed) (2017) Lineis describere. Sette seminari tra rappresentazione e formazione, Libria, Melfi

Luigini A (2019) Four dichotomies on digital environments between art, heritage and education: opening address. In: Luigini A (ed) proceedings of the 1st international and interdisciplinary conference on digital environments for education, arts and heritage. EARTH 2018. Springer, Cham, pp 3–12

Luigini A, Brusaporci S, Vattano S, Tata A (2019) 3D digital models for a widespread museum: the Renon's "bauernhöfe". Int Arch Photogramm Remote Sens Spat Inf Sci XLII-2/W9:447–453

Maiezza P (2019) Ricostruendo in digitale. metodi e modelli per i beni architettonici. PUBLICA. Sharing Knowledge, London

Moretti M (1971) Architettura medievale in Abruzzo (dal VI al XVI secolo). De Luca, Roma

Moretti M (1972a) Restauri d'Abruzzo (1966–1972). De Luca, Roma

Moretti M (1972b) Collemaggio. De Luca, Roma

Russo M, Menconero S, Baglioni L (2019) Parametric surfaces for augmented architecture representation. Int Arch Photogramm Remote Sens Spat Inf Sci XLII-2/W9:671–678

Spagnesi G (1984) Autonomia della Storia dell'architettura. In: Spagnesi G (ed) Storia e restauro dell'architettura. Istituto della Enciclopedia Italiana, Roma, pp 7–10

Steuer J (1992) Defining virtual reality: dimensions determining telepresence. J Commun 42(4):73–93
Sutherland IE (1965) The ultimate display. In: Proceedings of IFIP congress, pp 506–508
Trizio I, Brusaporci S, Luigini A, Ruggieri A, Basso A, Maiezza P, Tata A, Giannangeli A (2019) Experiencing the inaccessible. A framework for virtual interpretation and visualization of remote, risky or restricted access heritage places. Int Arch Photogramm Remote Sens Spatial Inf Sci XLII-2/W15:1171–1178
Torlontano R (ed) (2010) Abruzzo Il barocco negato. Aspetti dell'arte del Seicento e Settecento. De Luca, Roma
UNESCO (2003) Charter for the preservation of the Digital Heritage

Virtual Reality for Cultural Heritage: New Levels of Computer-Generated Simulation of a Unesco World Heritage Site

F. Banfi and C. M. Bolognesi

Abstract This research proposes a generative method capable of improving the interoperability levels of HBIM models for the generation of immersive environments based on the main game engines such as Unreal Engine. Thanks to advanced modelling techniques based on Non-Uniform Rational Basis-Splines (NURBS) algorithms, able to faithfully represent the surveyed artefact and an open-source visual scripting language for gameplay (blueprint for game engine), it was possible to implement an VR project of one of the most representative Unesco World Heritage Site of Lombard architecture: Santa Maria Delle Grazie in Milan (Italy), according to the Cloister of Dead part. In particular, thanks to HBIM and VR integration, the proposed method brings to light the intangible values of the historical monument, handing down the historical phases and the memory to future generations that have followed over the centuries from the first construction of the convent complex, to the reconstruction that took place after the bombing of the Second World War. Users, through an immersive path, can discover the transformations that took place over the centuries, which are gradually going slowly lost in our common memory. The virtual interactive reconstruction concerns specifically the project carried out and then demolished of the access to the convent from via Sassi and the transformations linked to the Cloister of the Dead before and after the bombing of 1943. Thanks to new levels of interactivity and state-of-the-art technologies, the VR project in Santa Maria Delle Grazie, which is being developed for different devices and platforms (mobile, desktop and Oculus Rift), aims to increase historical and cultural awareness of different types of users such as professionals in the construction sector and virtual tourists, thus becoming a useful dissemination tool for in-depth research carried out in recent years.

F. Banfi · C. M. Bolognesi (✉)
Department of Architecture, Built Environment and Construction Engineering, Politecnico di Milano, Via Ponzio 31, Milan 20133, Italy
e-mail: cecilia.bolognesi@polimi.it

F. Banfi
e-mail: fabrizio.banfi@polimi.it

C. Bolognesi and D. Villa (eds.), *From Building Information Modelling to Mixed Reality*, Springer Tracts in Civil Engineering,
https://doi.org/10.1007/978-3-030-49278-6_4

Keywords HBIM · Scan-to-BIM-to-VR · Virtual Reality · Santa Maria delle Grazie · Unesco World Heritage Site

1 The Origins of the Complex; Its Transformation, Until the Surveys of Agnolodomenico Pica and Piero Portaluppi

Probably count Gaspare Vimercati, captain of the Sforza troops, gift the land to a group of Dominican friars in 1460 to set the first stone of the convent in 1463 by a pre-existing military camp of his property. Later on Francesco Sforza, Galeazzo Maria and Ludovico il Moro from the Sforza family made the first donations and intervened in the new construction.

Duke Ludovico, (after a conspicuous donation in 1492) declared that his family had enlarged the main chapel and even built the whole church. During the first years, the new buildings of the old wooden military camp and the new buildings (already given to the friars and lasted till 1497) coexisted. After the lodging of the friars in 1463, the built continued with the construction of the church, next to a previous existing chapel, dedicated to the Virgin Mary. The first convent was completed in 1469, the church in 1490. The design of the first plan was the result of Gaspare Vimercati and Guiniforte Solari's work; its morphology was surely influenced by the presence of the military camp and by the position of the chapel of the Virgin Mary, already in place (Fig. 1).

The rules were typical of a Dominican convent, reflecting the necessity of a self-sufficient organization: spaces for praying, producing, living and resting. Different cloisters with different functions were built by the church, with different functions, such as a vegetable garden by the pharmacy and a cloister to bury the dead friars.

Some constraints were given by the lying of Porta Vercellina and the previous Chapel of the Holy Virgin; the influence was strong and durable; owing to the position of the Holy Virgin chapel the growth of the convent was strictly prolonged to the north and the lying of the Cloister of the Dead also (Bertelli et al. 1983).

In this paper the historical events that linked the development of the convent to the presence of Donato Bramante and Leonardo da Vinci are overlooked, included the decision of Ludovico il Moro to rebuild part of the complex.

In these years Leonardo da Vinci received the task of painting the Last Supper on a wall of the Refectory and the presence of this iconic painting influenced the whole history of the convent. The so-called Old Sacristy and the Cloister of the Frogs were finished before the beginning of the sixteenth century (but after the fall of the Sforza Family), while other spaces were later included, filling the gap between the cloister and the sacristy (Zanzottera 2015).

Fig. 1 The plan of the complex (*Source* A. Pica, P. Portaluppi) (**a**), Basilica of santa maria delle grazie (**b**), internal views of the church (**c**, **d**) the refectory and the last supper (**e**), the main cloisters (**f**, **g**, **h**)

1.1 The Cloister of Dead

The focus of this paper regards one of the bigger cloisters, the Cloister of Dead. After its construction well described by Armando Bruschi (Bertelli et al. 1983) it suffered from different minor interventions, some of them described in a set of photographs collected before the bomb of 1943. Its importance is linked to the history of the morphology of the convent.

The construction of the ehole convent complex must have probably started from the wing of the Cloister of Dead, close to the old Chapel of the Holy Virgin, first of all with the Chapter room, covered with vaults; the position of its opening into the cloister forming the typical design of the door flanked by two mullioned windows was transmitted to the mendicant orders with Cistercian monastic architecture, also Lombard. The Cloister of the Dead is formed by three built-arms: the one with the Chapel of the Virgin Mary, the chapter room, the one with the cells and the one with the refectory. With the side of the church, they form a squared space with columned porches whose principal walls will be able to further extend becoming the line structuring of most parts of the convent.

It may, therefore, be that the position of the main walls in the Cloister of the Dead, which delimit the building with the cells and the one with the refectory towards the outside, was suggested or made obligatory by the pre-existing buildings; this gives the importance to this cloister as an inner heart of the whole construction and could explain the specific lying of the refectory if considering the anomalous east and north side of the cloister; this great room has a perfect proportion of 4 modules with a width of this building bigger than the wing with the chapel and the Chapter. The entrance is considered in the middle of its long side (Fig. 2).

Fig. 2 The cloister of the dead and its architectural details in its actual shape

With this approach, the cloister of the Dead is almost a squared Cloister of almost 30 m each side, while elevation character among its sides is quite different. The first side built was the chapter one; to the north cells on the same length (destroyed in 1943) there were six spans instead of eight trying to correspond internal walls; in this way, the proportions established in the design of the eastern front were deformed here, without any visual concern, widening the centre distances of the columns and lowering the arches. Six arches were also attached to the western side of the Refectory. The fourth side of the portico, attached to the church (perhaps built last and also now destroyed), abandoned the articulation of six spans, changing them to five precisely to match the axes of the columns with the walls of separation of the chapels, without worrying of any irregularity.

In this way, the appearance of this fourth portico differed strongly in a particular way from that of the eastern side towards the chapel and the Chapter. While the eastern and northern arm even if with different heights comprised the cells and the library respectively on the upper floor, the western arm towards the Refectory and the southern arm towards the church, were originally as high as the porch alone. For those who entered the convent, the eastern side of the cloister, the most "proportionate" and regular with its eight-round arches, was the one that, despite an oblique view, appeared first and almost frontally.

1.2 After Renaissance

As many other religious institutions, the more recent history of this complex concerns the suppression of the convent in May 1799, when the building was transformed into a military barrack and the friars were shifted to other convents. Beside this in the middle of the eighteenth century, some urban transformations allowed a better view of the church and convent itself: during the years many buildings had been placed next to the church and their demolition, due to the opening of two new roads, offered better visibility to the apse.

Between the end of eighteenth century and the beginning of 19th, some restoration works were conducted in many areas of it, with the support of Luca Beltrami, who was specifically involved in the restoration of the Refectory. Due to a mixed-use, still divided between laic and religious, the convent underwent several transformations. The Regional Monuments Conservation Office had its seat there (Bascapè and Mezzanotte 1968) while in 1924 the Dominican friars regained possession of the courtyard called Cloister of the Prior and of the entire first floor. In 1929, the whole convent went back to the Dominican order with the exception of the Refectory containing the Last Supper, that remained a State property.

The main document that provides a first summary of the history of Santa Maria delle Grazie, as well as of its development over the centuries, is the book written in 1937 (in occasion of the renovation of the monument started in the 1930s) by Pica and Portaluppi (1938).

The volume contains previous historical researches that had been carried out by reliable scholars in the architectural field, such as Luca Beltrami, but also by friars belonging to the Dominican order, such as Fra Gattico (2004). It describes the origin of the complex, validated by the later writing of Bertelli et al. (1983) at the beginning of the 80s. Considering the complicated history of the monument, this is the most reliable document before the bombing of 1943 and gives the most complete overview of its morphology, giving the opportunity to understand its history (Fig. 3).

2 The Research Method

One of the main key elements for the development of the VR project of the Basilica of Santa Maria delle Grazie was the integration of different advanced modelling techniques able to interact with one of the most widespread applications for video game development: Unreal Engine 4 as already developed in a previous research Bolognesi and Aiello (2019).

The proposed workflow, besides allowing to inherit complex models developed in any 3D modelling software with high levels of interoperability, has allowed the authors to 'give life' to three distinct historical phases before the bombing of 1943 of a single part of the monument and associate useful contents, thus increasing the informative value of the virtual experience. Starting from the assumption that one of the main advantages of VR is the creation of an interactive environment capable of digitally representing the examined artefact, following the example of previous VR projects in this specific field of application (Ioannides et al. 2017) the main steps shown in Fig. 4 have been addressed.

2.1 The First Phase: 'Data Collection'

Data collection phase allowed to deepen the historical heritage of the basilica, analysing the various historical phases prior to the bombing suffered by the basilica during the Second World War in 1943. As is well know, VR allows us to digitally reconstruct and discover some areas of historic buildings forbidden to the public for a number of reasons such as maintenance works, limited use of certain areas to best preserve relevant parts of the building and particular parts of the building reserved for everyday life of the ecclesiastical body. Thanks to the in-depth study of a huge quantity of historical records during the last years, the main objective of this first phase was to reconstruct the historical background of the whole complex, thus facilitating its diffusion and observations among the various users (virtual tourist, students, professionals).

The determination of the various historical phases, as anticipated in the previous paragraph, is based on an in-depth study of historical documentation, scientific studies, restoration treatises and technical drawings made in the previous century,

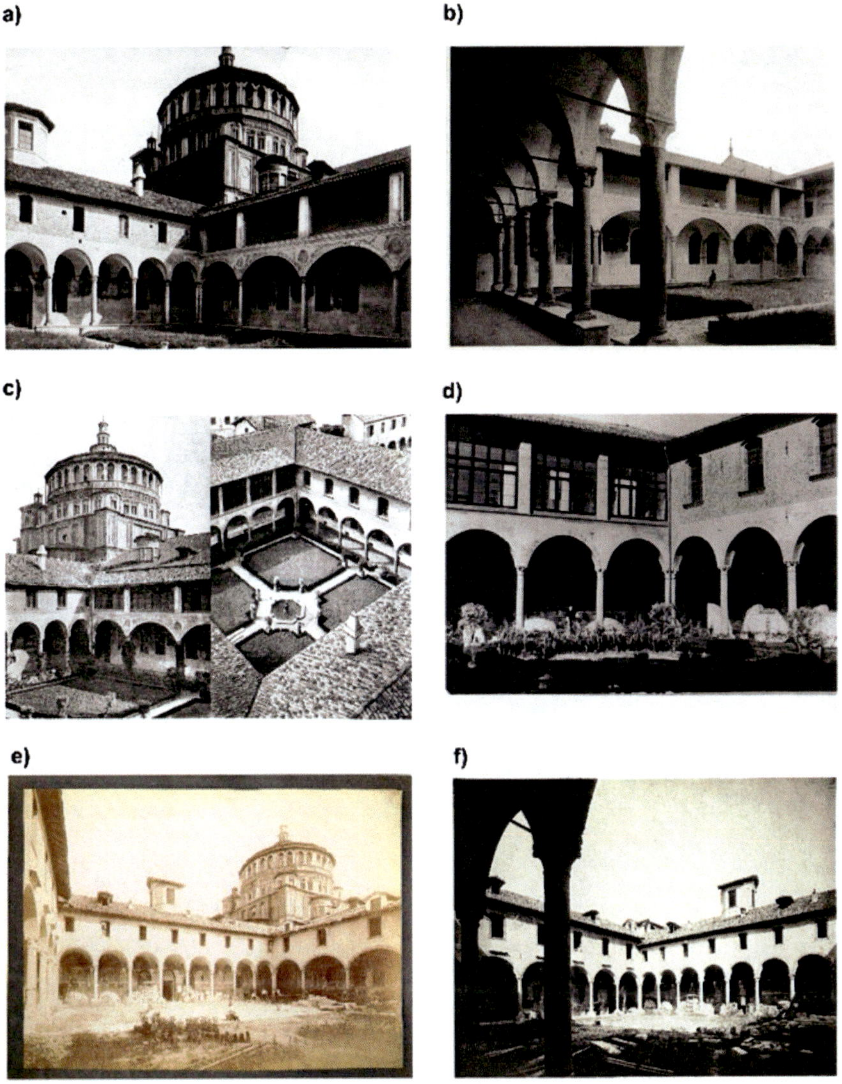

Fig. 3 Historical records of the cloister of the Dead in different temporary periods before the bombing by the Allies on August 16, 1943. *Source* **a-c** da A. Bruschi, in Santa Maria delle Grazie in Milano, 1983 Banca Popolare di Milano **b-d-e-f** Soprintendenza Archeologia, Belle Arti e Paesaggio, Milano

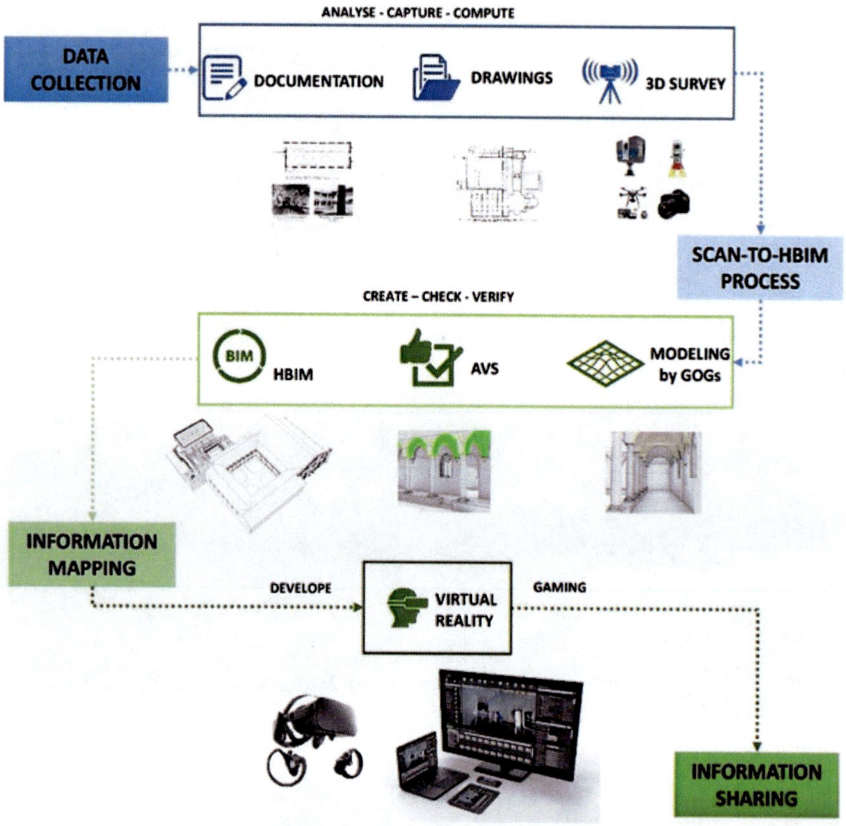

Fig. 4 The digital workflow applied to the VR project of Santa Maria delle Grazie

which have allowed authors to discover the morphology of the building during the centuries, its main historical configurations that took place in the years before the bombing.

In particular, this study makes explicit and compares three historical phases starting from the year 1937, with the main purpose of digitally telling the typological and morphological differences of the main cloister of the church (also known as the Cloister of the Dead) in an interactive virtual environment. In addition to increasing the constructive and geometric awareness of the basilica was to gather a series of information with the aim of validating with certainty the VR model corresponding to each historical phase and improving the level of content of the immersive experience itself.

2.2 *The Second Phase: The Scan-to-BIM Process*

In recent years a large number of studies and projects in the field of HBIM and VR have allowed the creation of faithful reproductions of a large number of heritage buildings, maximizing the '*value of the measure*' and the levels of information of the surveyed building (Banfi 2019; Graham et al. 2019; Arayici 2008; Antonopoulou and Bryan 2019).

The main objective of the Scan-to-BIM process is to generate detailed models corresponding to reality from different types of digital data. As is well known, today, instruments such as laser scanners, total stations and digital photogrammetry allow us to collect, store and analyze a large amount of data such as point clouds, accurate measurement, geodetic networks, and orthophotos.

These data collected during the first phase, allowed to provide the appropriate basis for the virtual reconstruction process.

2.3 *The Third and Fourth Phases (Information Mapping and Information Sharing) for VR Projects*

Information mapping and sharing phases have considered a specific selection of historical information relating to the three historical phases identified by the studies carried out with the aim of developing a virtual storytelling for multiple devices such as the VR headsets, portable devices (tablet, mobile phone and laptop) and pc. Starting from the assumption that every device and software requires a high level of knowledge in the field of HBIM modelling, thanks to this open-source application, the main research objective was to maximize the level of content in a virtual environment and to create the most sustainable HBIM-to-VR digital process possible even for future users.

For this reason, the UE4 VR project of the basilica had to provide an open approach able to give the possibility to enrich the VR experience and the virtual storytelling to users and studies successes to the one proposed (Fig. 5).

As shown in the following paragraphs the VR project of the Basilica of Santa Maria delle Grazie will take different forms based on the software and devices used with a single final purpose: to increase the historical and cultural awareness of one of the main Unesco sites in the city of Milan.

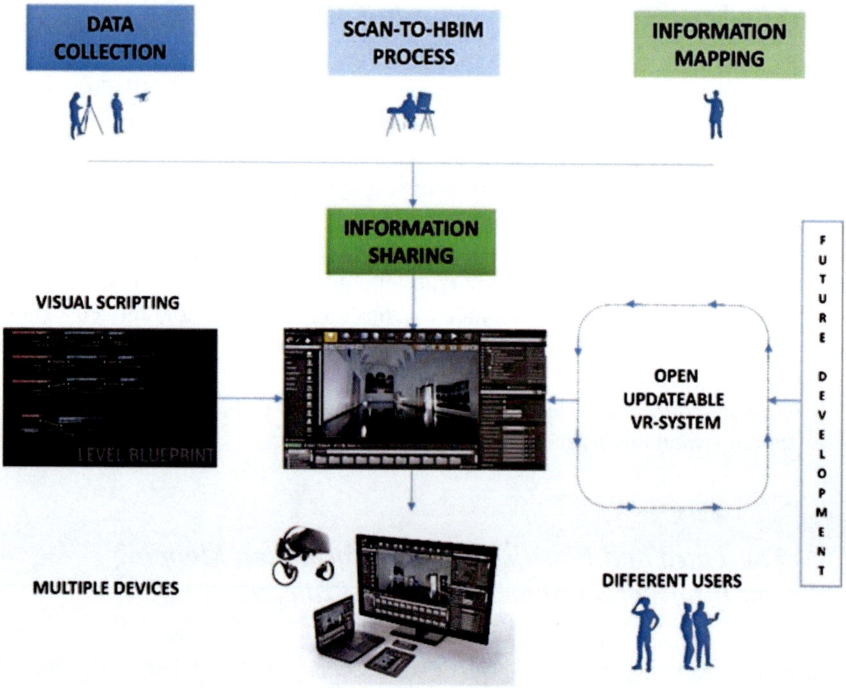

Fig. 5 The holistic approach based on a open visual scripting (blueprint of UE4) allow future users to improve the VR system with new type of contents and orient it for multiple devices

3 From 3D Drawing and Digital Modelling to Virtual Reality (VR)

Considering the previous studies Bolognesi and Aiello (2019), this research proposes a sustainable digital process for the creation of a VR project able to improve the information sharing of one of the most important monuments of Milan. As anticipated in the previous paragraph, HBIM makes it possible to improve different types of analysis thanks to the targeted use of complex models capable of faithfully representing reality with novel frades of generation. The main VR applications such as Unreal Engine and Unity provide a type of mesh-based modelling. Mesh modelling, as is well known, involves a series of pros and cons. Considering the world of digital photogrammetry (Lerma et al. 2010; Barba et al. 2019), thanks to simple sets of photographs it is possible to move from a two-dimensional representation to a three-dimensional geometric representation composed of a large number of polygons, precisely the meshes.

Thanks to the generation of models from point clouds, it is possible to create texturized geometric entities with high-resolution orthophotos. On the other hand, one of the main cons of this type of modelling is the realization of a geometric model and not of a parametric HBIM model able to communicate different levels of information such as the physical and mechanical characteristics of the materials, historical phases, wall stratigraphy etc.

In this specific field, the integrated application of GOGs and NURBS algorithms (Banfi 2017; Piegl and Tiller 2012) has made it possible to generate the model of the Basilica of Santa Maria delle Grazie by maximizing the benefits deriving from 3D survey for the Cloister of Frogs and of the Priore, and the extraction of geometric primitives from historical drawings such as elevations, plans, detailed sections for the Cloister of Dead here rapresented. A mathematical model able to interact with three different historical phases has been prepared. Thanks to the sharing of a unique 3D project it has been possible to insert 2D and 3D construction details in a georeferenced environment with the aim of fully understanding the construction technique and fixing possible interferences between one historical phase and another. Figure 6 shows the various steps in detail, from the generation of the model to the VR development environment. The main steps were:

- use of different types of data such as point clouds and 2D technical details (plans, elevations and sections)
- generation of geometric models using NURBS and GOGs interpolation algorithms
- verify the main interferences between a historical phase and the other, outlining a sure road to take for the generation of the models;
- texturing of NURBS models using the main mapping and software techniques such as McNeel Rhinoceros with UV mapping tool and Spotlight with its add-in Zbrush; paying particular attention to the direction of the normal of complex NURBS surfaces
- import of the NURBS model in Unreal engine with a workflow based on the main exchange formats (obj and fbx) able to accurately transmit the geometric and material characteristics of each single element created; for this step, satisfactory results have been found also through the use of Datasmith (new add-in for UE4), which allowed to bypass the definition of the export schemes of the two previous formats and directly use the proprietary files such as the 3 dm format (Mc Neel Rhinoceros) and the various CAD and BIM formats in a completely open logic.

3.1 The Development of the VR Project of Santa Maria Delle Grazie

Before the IT development of the Santa Maria delle Grazie VR project, it was necessary to define the "how" to transmit the wealth of studies and research done in recent years. For a correct setting of a VR environment, it is necessary to identify and decide the various LOD that the virtual user must discover when he explores the model. In recent years, some studies on proxemics, perception of virtual space and above all cognitive learning have highlighted key features that can greatly influence the

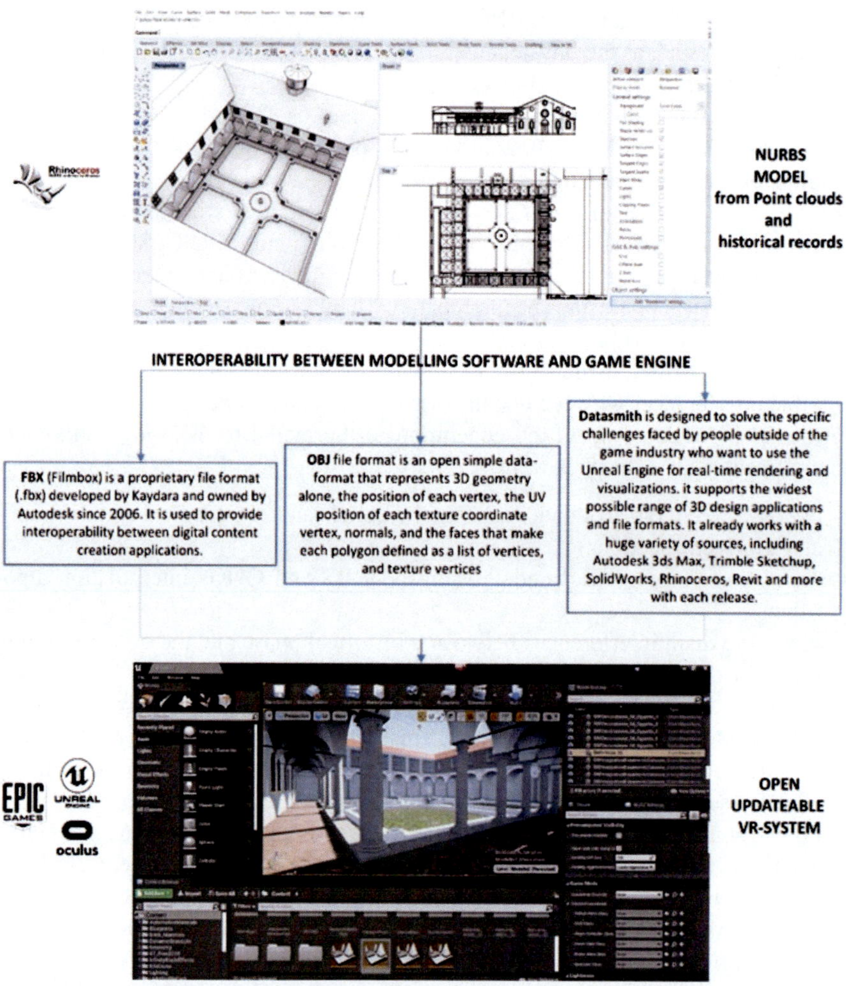

Fig. 6 The three import options of the HBIM model for VR environment and game engine

virtual experience itself (Michael and Chen 2005; Guidi et al. 2007; Westhoven and Alexander 2015; Hayden 2018a, b). Independently from the LOI and LOD achieved in the previous stage, these studies underline the importance of serious games for their ability to fulfil cognitive learning outcomes and motivate users. It has been tried and proven through these studies, serious games are designed for a general audience and often require a user-friendly interface tailored to learning objectives. Notable among studies of selecting game engines for serious games is the work of Petridis et al. (2010) and Pybus (2019), which identifies several key criteria for comparison that are still useful despite changes in technology: audiovisual fidelity, functional fidelity, composability, accessibility, networking, and heterogeneity. Considering all these criteria, for a correct setting of the basilica storytelling and its three historical

Fig. 7 The 'lobby' of the VR project: the starting point of the virtual immersive experience allow users (through the VR avatar) to select different type of data (on the left) and choice the three time gates corresponding to each historical phases analysed (on the right)

phases, it was necessary to create a starting point where the user can begin to explore and learn key information of the basilica and decide how to interact with the various VR objects and the three historical phases identified. This k-concept was also useful for all those users who are not able to visit the monument directly and to better understand general information such as the context, location and history of the basilica. For this reason, a 'lobby' has been created which takes up the studies done by the restoration of the refectory where Leonardo's last supper is located (Fig. 7).

3.2 *The Five Rules Applied to the VR Project*

The choice to undertake the modelling of such a famous environment was given by the authors' desire to create three temporal links (one for each historical phase) to the cloister of the Dead (Fig. 8). Starting from the assumption that the storytelling of a building of such a historical and cultural importance of the city of Milan cannot be left to chance, it was useful to identify five basic rules to be able to tell successfully the articulation of the virtual experience, its storytelling and the three identified historical phases.

Linear and engaging story: identification of the protagonist of the narration, periods of the story told, identification of the elements of interest for a target that can go from the historical to occasional virtual users;

Reliability of the model and of the sources used: use of qualified sources that are as objective as possible and that can confirm the assertions made in the narrative and of the models created;

Identification of the best VR devices and software: the creation of a serious game must include a high level of authenticity, realism and persuasive effectiveness. To obtain these results, the main development applications such as Unreal Engine

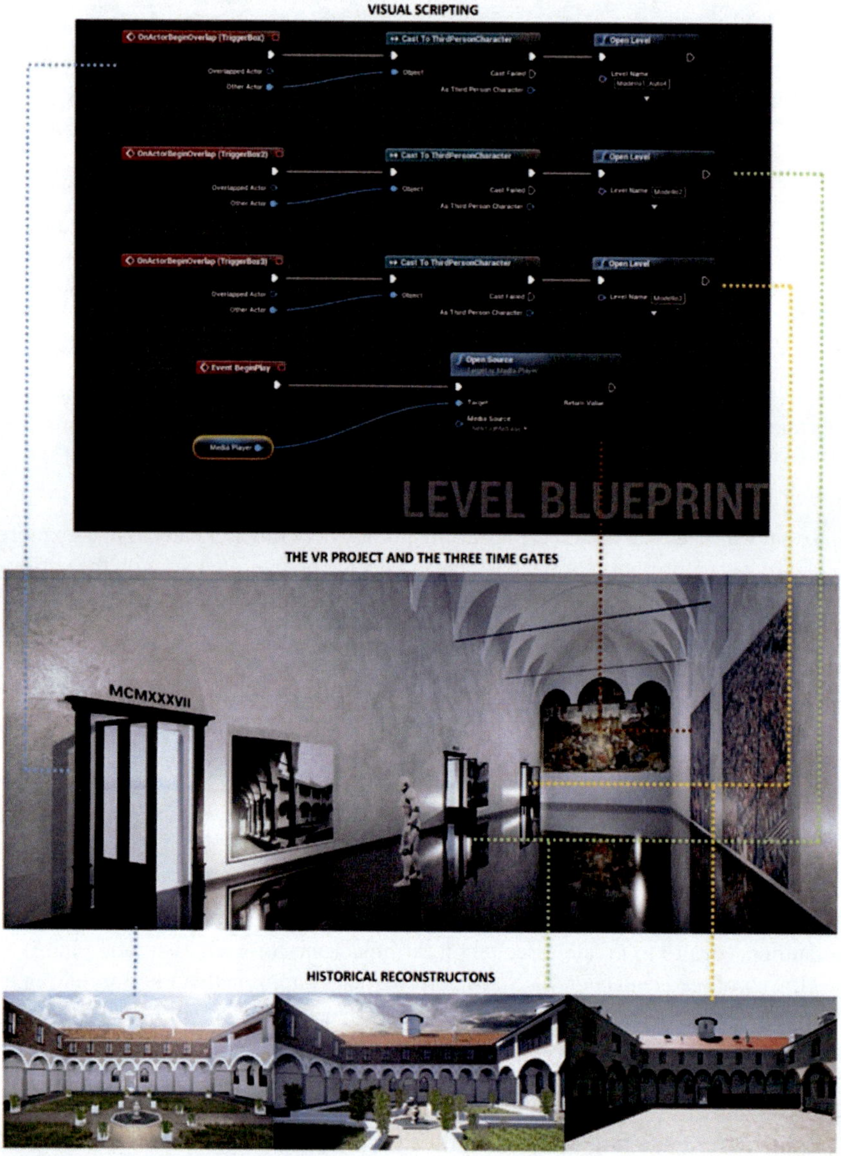

Fig. 8 The main blueprints developed for the VR project of Santa Maria delle Grazie and the main three levels developed

and Unity and the main VR headsets and devices such as PlayStation VR, Oculus Rift S, Samsung Gear VR and Oculus Quest were considered. Unreal Engine was preferred thanks to its visual fidelity, simplicity of scripting, learning curve, and the availability of online resources including documentation and community support; while Oculus Rift S was chosen (Oculus/Developers 2019; Addison and Gaiani 2000; Remondino and Stylianidis 2016; Zancheti 2002) thanks to the latest technological developments that have improved the optics, improved the touch controllers' accuracy and reduced the hardware needed to immerse themselves directly in the virtual experience, eliminating the turrets included in the standard Oculus rift version.

Control of the size of the VR project in terms of bytes and polygons: the size of the VR project depends on the number of mesh polygons obtained from the transformation process of the NURBS models into static meshes of unreal Engine. Thanks to the definition of the exchange format and the use of the Mc Neel Rhinoceros software it was possible to decide the number of polygons of every single element before importing it into the VR engine. This step allowed the modellers to keep the overall size under control, favouring the app package function for different devices such as mobile phone and tablets which require a limited number of polygons.

Guarantee future developments and integrations through a VR project based on open development logic: the advantage of having developed the VR project with Unreal Engine as well as orienting the app for multiple devices, was to develop algorithms and information codes through visual scripting. Thanks to the flexibility of this programming language it has been possible to update the virtual experience and the storytelling with past studies, thus ensuring a flexible and updatable implementation at any time.

4 Discussion and Results. HBIM to an Open VR Project: Data Enrichment and a New Level of Interoperability

In recent years, the generation of complex HBIM models capable of representing the detected reality has been oriented towards disciplines such as restoration, geomatics, structural simulations, maintenance site and facility management. Many studies have dealt with and resolved differently the limits imposed by BIM platforms, implementing methods capable of managing both the latest generation of relevant tools and advanced modelling techniques, the latter not always able to facilitate the generation of complex historic buildings.

A known limitation of the users is the ability to effectively manage, in order to interact satisfactorily, complex three-dimensional models characterized by huge amounts of data, such as those generated by the most advanced technologies today, due to the limitations resulting. It happens, therefore, that information models of considerable dimensions, such as those generated by the 3D survey, are subject to discretization and therefore a choice to reduce the information content, in particular when the purpose becomes the visualization through common devices.

For those reasons, the technological development, operational and modelling aspects have been investigated with the main objective of demonstrating how the proposed method can be useful and sustainable for all those fields of application related to the built heritage, not only to the fields of restoration and safeguarding the good. This study has shown how the new paradigm of the utility of HBIM models can be addressed to new tools and devices capable of increasing the level of information and transmissibility between different types of users, from professionals to tourists and virtual students, through an open and sustainable approach.

In particular, the choice of modelling techniques, the software and the devices used were dictated by the fact of making immersion in digital worlds (made up of VR objects) as user-friendly as possible, moving from static information models to interactive objects and scenarios characterized from its own life, capable of communicating data and information that cannot be associated with traditional HBIM projects. The description of the technologies used, as well as the procedures adopted, should always be related to the characteristics of the target audience considered.

The use of applications such as Rhinoceros V6 and Unreal Engine allow to model with extreme precision deformation states detected through the three-dimensional survey of the actual state of the building; these software solutions have considered from the authors as the best choice to generate models able to interact with different platforms, devices and users. As is well known, Rhinoceros V6 is considered the best digital model converter, including all possible exchange formats within it. Thanks to the choice of formats such as .fbx and .obj, it was possible to use a VR platform based on open development logics, importing complex NURBS model in an open logic.

Furthermore, it is also considered that the presence of developing protocols and procedures shared online by Epic Games for the Unreal Engine software can support both expert and non-expert in creating new and innovative digital storytelling, following their development and storytelling creativity. As a result, semantic enrichment becomes the added value of the proposed VR project, which can tell the story of the building and pass on the historical and cultural awareness accumulated in recent centuries. Informative panels, interactive VR objects, historical phases, descriptions, videos, audios, pictures and other types of multimedia data are therefore the 'key elements' for improving the transmission of information within a virtual model. As briefly mentioned, the intention of using software based on open development logics also opens the door to possible future implementations.

The semantic enrichment of the VR project consequently has no limits, thus depending solely on the will of the creators and future developers to increase the content and the level of information. This open approach has also favoured the development of 'temporal doors' capable of increasing the educational value of the VR project, through which the virtual tourist can immerse himself and discover different historical periods, discovering restorations and architectural differences that have alternated in a specific time period.

This study must, therefore, be seen as the first attempt at digital development and semantic enrichment of the VR project of the basilica of Santa Maria Delle Grazie.

Many other enrichments can be made, following the example outlined by this first phase of development.

5 Future Research, Development Perspective and Conclusion

Today's people are more tech-savvy than any other generation. Thanks to their multiple advantages, gaming technology and VR have been considered very good ways to improve the knowledge of heritage sites.

VR and digital modelling allow users to recreate places normally not accessible to people, to allow realistic immersion in digital environments that normally could not be explored, to share a huge quantity of contents through a digital 3D reconstruction and to improve the level of interactivity and knowledge of different types of users, from professionals to virtual tourists and students.

For all these benefits, one of the main research goals was the integration of the latest modelling techniques, 3D survey and in-depth studies of the Basilica of S. Maria Delle Grazie to create an educational VR project variable and extendable both in terms of contents, interactivity and creativity, increasing the historical and cultural awareness of one of the most famous UNESCO monuments of North Italy.

Furthermore, thanks to its open logic, the proposed method will enable users to extend the storytelling of the church, introducing new historical records and multimedia data, discovering new types of devices and software.

Finally, new development perspectives will able to be based on the proposed digital workflow for other types of heritage sites, considering it a possible base for their case studies.

References

Addison AC, Gaiani M (2000) Virtualized architectural heritage: New tools and techniques. IEEE Multimedia 7(2):26–31

Antonopoulou S, Bryan P (2019) Historic England BIM for heritage: developing a historic building information model. Historic England, Swindon. Retrieved 10 Mar 2019

Arayici Y (2008) Towards building information modelling for existing structures. Struct Surv 26:210–222

Banfi F (2017) BIM orientation: grades of generation and information for different type of analysis and management process. Int Arch Photogramm Remote Sens Spat Inf Sci, 42(2/W5):57–64

Banfi F (2019) HBIM Generation: extending geometric primitives and bim modelling tools for heritage structures and complex vaulted systems. ISPRS - Int Arch Photogramm Remote Sens Spat Inf Sci, 42(2/W15):139–148

Barba S, Barbarella M, Di Benedetto A, Fiani M, Limongiello M (2019) Quality assessment of Uav photogrammetric archaeological survey. Int Arch Photogramm Remote Sens Spat Inf Sci

Bascapè C, Mezzanotte P (1968) Milano nell'arte e nella storia. Carlo Bestetti, Milano

Bertelli C, Bora G, Bruschi A, Caccin AM, Ciati B, Giordano L, Gremmo L, Mulazzani G, Pesenti PZ (1983) Santa Maria delle Grazie. Cinisello Balsamo, Amilcare Pizzi Arti Grafiche. Banca, Popolare Di Milano

Bolognesi C, Aiello D (2019) From digital survey to a virtual tale: virtual reconstruction of the convent of Santa Maria delle Grazie in Milan. In: Bolognesi CM, Santagati C (eds) Impact of industry 4.0 on architecture and cultural heritage. Cultural computing series. Springer, Berlin, pp 49–75

Bolognesi C, Aiello D (2019) The secrets of S. Maria delle Grazie: virtual fruition of an iconic milanese architecture. Int Arch Photogramm Remote Sens Spat Inf Sci

Gattico G (2004) Descrizione succinta e vera delle cose spettanti alla chiesa e convento di Santa Maria delle Grazie e di Santa Maria della Rosa e suo luogo, et altre loro aderenze in Milano dell'Ordine de'predicatori con due tavole in fine. Ente raccolta Vinciana, Milano

Graham K, Pybus C, Arellano N, Doherty J, Chow L, Fai S, Grun T (2019) Defining geometry levels for optimizing BIM for VR: insights from traditional architectural media. Technol Arch Des 3(2):234–244

Guidi G, Frischer B, Lucenti I (2007) Rome reborn—virtualizing the ancient imperial Rome. In: Workshop on 3D virtual reconstruction and visualization of complex architectures

Hayden S (2018) Award-winning VR space experience 'SPHERES' lands on rift. Road to VR, 14 Nov 2018

Hayden S (2018) David attenborough's VR experience 'hold the world' lands on rift. Road to VR, 03 Nov 2018

Ioannides M, Magnenat-Thalmann N, Papagiannakis G (2017) Mixed reality and gamification for cultural heritage. Springer, Cham

Lerma JL, Navarro S, Cabrelles M, Villaverde V (2010) Terrestrial laser scanning and close range photogrammetry for 3D archaeological documentation: the Upper Palaeolithic Cave of Parpalló as a case study. J Archaeol Sci 37(3):499–507

Michael DR, Chen SL (2005) Serious games: games that educate, train, and inform. Muska & Lipman/Premier-Trade

Oculus/Developers (2019) Guidelines for VR performance optimization. Oculus documentation. Available: https://developer.oculus.com/documentation/pcsdk/latest/concepts/dg-performanceguidelines/. Accessed 1 Jan 2019

Petridis P, Dunwell I, de Freitas S, Panzoli D (2010) An engine selection methodology for high fidelity serious games. In: 2010 second international conference on games and virtual worlds for serious applications, pp 27–34

Pica A, Portaluppi P (1938) Le Grazie. Casa Editrice Mediterranea, Roma

Piegl L, Tiller W (2012) The NURBS book. Springer, Cham

Pybus C (2019) New tools for cultural heritage tourism: accessible virtual reality for Milan's Basilica Sant'ambrogio. Int Arch Photogramm Remote Sens and Spat Inf Sci

Pybus C, Graham K, Doherty J, Arellano N, Fai S (2019) New realities for Canada's parliament: a workflow for preparing heritage BIM for game engines and virtual reality. Int Arch Photogramm Remote Sens Spat Inf Sci

Remondino F, Stylianidis E (2016) 3D recording, documentation and management of cultural heritage, vol 1. Whittles, Dunbeath, UK

Westhoven M, Alexander T (2015) Towards a structured selection of game engines for virtual environments. In: Virtual, augmented and mixed reality, pp 142–152

Zancheti SM (2002) Values, built heritage and cyberspace. Mus Int 54(3):19–28

Zanzottera F (2015) Il chiostrino di Santa Maria delle Grazie. SIRBeC ARL-LMD80-00362

VR and the Perception of the Space: A Sensorial Labyrinth Inspired by Giambattista Piranesi's *Carceri D'Invenzione*

Simona Calvagna, Federica Grasso, and Cettina Santagati

Abstract The Virtual Reality project called *Piranesi "beyond the real"* aims to help the understanding of the series of etchings *Carceri d'invenzione* (*Prisons of invention*) by Giovanni Battista Piranesi, exhibited in the Museum of Representation of DICAR (University of Catania), participating in the strategies of co-creation of cultural content implemented by the museum. The research begins with a careful investigation of the etchings, which reveals immense spaces where bridges, towers, and stairs seem to follow each other endlessly. The only things in between are wooden beams, ropes or instruments of torture, used by surreal giant men in the foreground or "human ants" in the distance. The spaces that inspire this project are intended as the dream of a young man because of the fever or the genius of his black mind. The virtual labyrinth is a journey designed from the sensations of the Prisons, an immersive experience in completely new environments. Anxiety, disorientation, and illusion are the focus, the goal is the awareness of them. Two levels, two different perceptions of space: the first is an immense and dark underground labyrinth, which sinks vertically towards infinity with flights of stairs, high walls, suspended or recessed platforms; the second is an open space, which extends towards the horizontal infinity, extremely bright. it is a conceptual labyrinth, which can be crossed in many ways, with endless exits, and an end. That is the place where the visitor can complete the ascent being aware of the places he crossed. The study of spatiality was also conducted through the aid of a 1:200 scale physical model, which allowed us to verify the project hypotheses.

Keywords Virtual reality · Labyrinth · Sensorial · Piranesi · Carceri d'invenzione · Museum storytelling

S. Calvagna · F. Grasso · C. Santagati (✉)
Department of Civil Engineering and Architecture, University of Catania, Catania, Italy
e-mail: cettina.santagati@dau.unict.it

S. Calvagna
e-mail: simona.calvagna@unict.it

F. Grasso
e-mail: f.grasso94.fg@gmail.com

C. Bolognesi and D. Villa (eds.), *From Building Information Modelling to Mixed Reality*, Springer Tracts in Civil Engineering,
https://doi.org/10.1007/978-3-030-49278-6_5

1 Introduction

The project intends to recreate in VR a *Piranesian space* in a contemporary key. It was born with the intention of increasing the understanding of Giambattista Piranesi's exhibition at Mura, Museum of Representation of the University of Catania. Virtual Reality (VR) has been chosen as a better method to achieve that purpose. Museums, in fact, are increasingly integrating VR in their exhibit because of the immersive experience they offer, which leads to a deeper comprehension of it.

The project has been developed with the support of the two laboratories of MuRa: R^3D (Laboratory of Survey, Representation and Digital Reconstruction) and SPrA (Instruments for the Architectural Project).[1] In this context, the project has been realized in both virtual and real way, as a VR project to be immersively experienced by using the visor and as a physical model to be seen from the exterior to better understanding the totality of the spaces created.

After the careful study and analysis of the life and works of Giovanni Battista Piranesi, the attention was focused on his revolutionary idea of 'space', so important in all the works he made. In particular, the study of the Prisons space was deeply analyzed through a better study of the atmospheres and figurative characters that Piranesi was able to transmit through the masterful control of the instruments of representation.

The first question made was: how could the intentions of Piranesi expressed inside the *Carceri d'invenzione* be better appreciated? And so, how can the visitor understand what inspired Piranesi in making those etchings?

The study of the *Carceri d'invenzione* led to identify the attributes of what the critics call a "Piranesian space": an immense labyrinthic space, where silence and darkness face chaos and light, where interior and exterior are strictly related, in a sequence of contrasts that lead to paradox. The visitor of that space is inexorably overcome by anxiety and needs to find a way out by exploring and, consequently, discovering the hidden from the evident (Fig. 1).

The space projected starts from these concepts that identify a *Piranesian space,* creating a new one in a contemporary vision: it is dominated by contrasts, sometimes defined by light over darkness and sometimes by darkness over light.

Space has been conceived as a virtual expansion of the museum and consists of two parts on two different and separate levels. It is a labyrinth, but no longer a classic one, defined by high walls, intersections and blind spots, but an infinite labyrinth. It becomes a maze, conceived as a desert, without landmarks and without corridors or tunnels. It is precisely the absence of these elements that makes it a maze, in which the search for a way out from an environment that seems always the same goes on forever.

---

[1]This project was realized as graduation thesis of Federica Grasso in Building Engineering and Architectural Construction at University of Catania, titled "Piranesi oltre il reale. Progetto di un labirinto sensoriale virtuale ispirato alle *Carceri d'invenzione*". The professors Simona Calvagna and Cettina Santagati were the thesis' mentors.

Disorientation and anxiety are the feelings that overwhelm the visitor while walking through the space. The visitor will start from the downer part and then arrive, through a journey of ascent and discovery of the meanders of a real underground labyrinth, to the top, where the labyrinth becomes abstract.

The end of the path marks the culmination of the ascent, where the visitor gets to the highest point, from which he can see the route made and become aware of how he got there and of the spaces he passed through. It is so a mental labyrinth, a psychological maze from where is impossible to escape because is nothing else than the world we are living in and where we are trapped in, or rather, our perception of it. After walking this path inside of the *Piranesian space*, the person who visits understands that he must find a way to live in it instead of a way out. It is a mental prison, that here is represented by a sensorial labyrinth whilst in *Carceri d'invenzione* by real prisons. The main aim that Piranesi wants to reach through this work is perhaps to lead the observer to understand this concept, to understand that the *Carceri* are nothing else than a reflection of real life.

The chapter is structured as follows: it begins with the description of the state of the art, dealing with VR today, virtual museums and the meaning of sensorial spaces in the project of architecture; then, objectives and methodology of the project will be shown, before to move on the case study, i.e. the Museum of Representation (MuRa), Giambattista Piranesi and the virtual sensorial labyrinth; and, finally, the conclusions will be drawn.

2 Background

2.1 VR Experiences and Virtual Museums

Virtual Reality was born with the 'experience theater' and the machine *Sensorama*, created by Morton Heilig in 1962, with the purpose to immerge the spectator into the movie by putting him inside of the action that was going on the screen through the senses of sight, hear, smell and tact. After that, the term began to be used to indicate the simulation of spaces or realistic situations where the user was invited to interact with the scene trough visors, data gloves and joysticks (Fig. 2).[2]

The concepts on which VR is based are tree: immersivity, interaction and connection of the user with the ambient. These features define the level of perception in virtual reality and allow to classify a space as believable.

VR headsets now in use have been developed from different companies, in detail the most used are *HTC Vive Pro, Oculus Rift S, Samsung New Gear VR,*

[2]The first visor, called *Damocles' sword*, was invented in 1968 by Ivan Sutherland and Bob Sproull and it was so heavy that it required a support structure. However, the visor that opened the era of virtual reality has been the *Aspen Movie Map*, a software invented by MIT in 1977 that let people walk virtually through the streets of Aspen, Colorado.

Fig. 1 Giovanni Battista Piranesi, Carceri d'invenzione, second edition, 1760, plate XIV. *Credits* Princeton University Art Museum

Google Daydream and *PlayStation VR*. There are also visors that can be used with smartphones, such as the *Google Cardboard*.

A visor need to satisfy few parameters to be used for VR, it needs: a field of view between 100° and 110°; a frame rate between 60 and 120 fps (frame per second); a gyroscope, an accelerometer and a magnetometer, in order to have a head tracking between 50 and 60 ms; a professional audio system; a system of eye tracking that allows to read the movement of the eye though infrared pointers.

Virtual Reality is mostly used for video games, although there are many fields of application. Its use is growing more and more also in the field of cultural heritage because it allows people to visit expositions and archeological sites in a more interactive way. Despite this, lots of museums managers limit the use of VR due to the inclination to alienate the visitor from the world. The most interesting applications in this field are the virtual visit to the digital reconstruction of monuments in their original state, or of sites that are inaccessible or hard to reach.

There are many software optimized to make VR projects. They are created to develop video games, but it is possible to use them also in the architecture field. The two most common open source software are *Unity* by *Unity Technologies* and *Unreal Engine* by *Epic Games*. Additionally, there are software created specifically

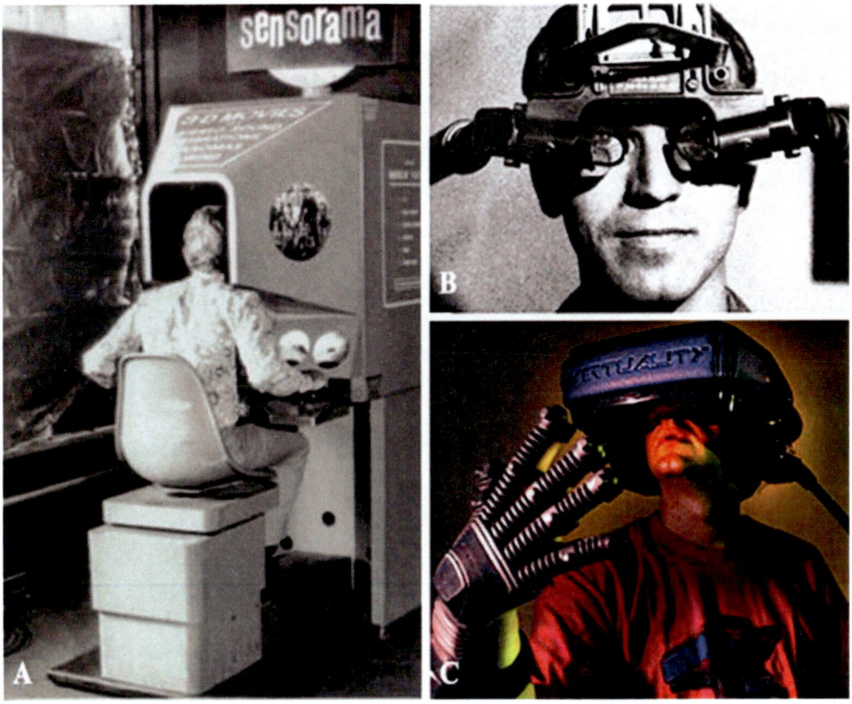

Fig. 2 **A** Sensorama, the first virtual reality machine; **B** Damocle's sword, the first virtual reality system with headset (so heavy it had to be supported by a structure); **C** Virtuality headset and glove. *Credits* A, B & C: Bianchini, Riccardo, "Quando i musei diventarono virtuali" in *Inexhibit*, 10/06/2016

for architecture visualization and design, e.g. *Eyecad VR* by *Digital Atoms. Unreal Engine* is the software used for the present project.

The concept of 'Museum' has changed substantially over time: from the traditional museum, seen as the planning of exhibitions of important works of art and objects inside a building, to the museum in a broader sense, seen as "a non-profit, permanent institution in the service of society and its development, open to the public, which acquires, conserves, researches, communicates and exhibits the tangible and intangible heritage of humanity and its environment for the purposes of education, study and enjoyment". (ICOM 2007) This definition is not considered any more appropriate to the contemporary museum. For this reason, during the ICOM's 25th General Conference, held in Kyoto (Japan) from 1 to 7 September 2019, a new definition has been proposed. This new definition hasn't been approved and it's still being discussed to ensure that it adapts as much as possible to the museum as it is today.

The idea of 'virtual museum' was born around the 1980s, when computers began to be commonly used. Museums didn't change many since the nineteenth century, they were no longer suitable for a society that developed so much. There was the need

Fig. 3 Digital installation at MuRa. *Credits* Museum of Representation

to adapt the museum reality to the new people's way to live. Information technology and virtual resources were the key to promote the culture inside museums (Figs. 3 and 4).

Initially 'digital museum' were created with the objective of speeding up the process of archiving, inventory and administration. Many museums also created their own CD-ROM with their virtual clone with the dual purpose of digital archiving and of better fruition of the works contained in the museum. In the following decade the Internet rapidly expanded, giving to museums the possibility to share their digital parts with the rest of the world. In the meantime, many museums started to have virtual parts, where the visitor could have an immersive experience, and sometimes a dedicated computer room for digital consulting, with the purpose of deepening the works or better understand some aspects of them.

The virtual museum has enormous potential: along with the traditional methods, it can help to restore importance and visibility to the museum's institution, creating "a dimension that allows us to experiment on the entire cultural heritage with cultural experiences and social interaction that are impossible in real-world institutions." (Galluzzi 2010).

Fig. 4 Virtual exploring of a 3D model from photogrammetric survey at MuRa. *Credits* Museum of Representation

2.2 *What Is a Sensorial Space? How to Project It?*

"Quality architecture to me is when a building manages to move me. […] One word for it is atmosphere. […] I enter a building, see a room, and—in the fraction of a second—have this feeling about it." (Zumthor 2006).

Atmosphere stays in the quality of the space, in the first emotional perception, in the man-things relationship, in the intensity and density of the space and in the relation between the objects or the materials inside of it. Each building gives a unique experiential perception, it generates its own sensations that will not ever be the same in another space. Walls, surfaces, columns, pillars, changing materials are defined by the tension between them. "Space is the essential mean of architecture. It is many things simultaneously: the voids and space around the architecture, the vastness of the landscape and the city, the intergalactic space of the universe. Space is something intrinsic and relational." (Holl 2004).

People, walking through the architecture, modify the space and have a perception of it that depends on those changes. For this reason, everyone can feel the atmosphere of a space in his personal way. It becomes "expression of moods in lived situations, which occur in habitual human action; guided by a poetic objective." (Pérez 2018). Furthermore, memories bind the person to a specific place, and it can influence the perception of other similar spaces, making them associate the feeling of the past place to the recent seen one.

The movement itself generate a connection between the guest and the architecture. The person moves inside the architecture, discovering its parts, its materials, modifying the plans that delimit the space during the movement. This process is called *parallax*, i.e. the movement of the object seen due to the movement of the observer. "Vertical or oblique movements through urban space multiply out experiences." (Holl 2004). Light also is part of this process: light generates movement and movement produces the perception of the space. "Architecture manifests itself in perception" says Holl (2004). Light, time, material and detail merge, becoming a 'whole'. They can't be perceived separately anymore.

There are 9 points that characterize sensoriality, emotional expressions and so the atmosphere of a building, according to the swiss architect Peter Zumthor, those are: the 'materials' that emanate their own quality from their relation; the 'sound', the noise of that site, the silence from where architecture is created; the 'body' of the building that you can touch; the 'temperature' of the building; the 'objects' that characterize the life and the presence of the residents; the relation 'between composure and seduction' in wandering within spaces; the 'tension between interior and exterior'; the 'levels of intimacy' that result in distance and proximity, in weight, volume of objects, in the relationship between their dimensions within the space; the 'light' that generates from darkness, defining the spaces (Fig. 5).

The project of a sensorial space starts from these principles, taking them as a reference in relation to the characteristics of the Piranesian spaces.

Fig. 5 Bruder Klaus Chapel, Peter Zumthor, Mechernich, Germany, 2007. http://sacrark.altervista.org/peter-zumthor-bruder-klaus-kapelle/

3 State of the Art

In museum context Virtual Reality is starting to become very used. In the following part are showed significant examples.

The Hunterian Museum at the University of Glasgow (UGLA) in 2017 started a 3-year research and innovation project, called *EMOTIVE Hunterian Museum digital storytelling about the Antonine Wall.* This project aims to use emotional storytelling, through the EMOTIVE tools, to change the traditional experience of museums and heritage sites. The EMOTIVE project offers different tools to create experiences for on-site or remote visitors, such as 'Interactive storytelling experiences for mobile devices' (an app for smartphones that guides the visitors around the museum though immersive narratives), 'Bring your experiences online' (an editor that allows to easily create a 360° virtual space from photographs and share it on the EMOTIVE's Web Experiencing System in order to be experienced off-site), and 'Bring objects to life' (the combination of a 3D printer that cast historical artefacts' replicas and the EMOTIVE's Mixed Reality Plugin for Unity that brings back the objects to their original glory using a VR headset).

Immersive VR/AR Museum Guide for the Miocene Site of Stetten, Austria is a project developed in 2017 for the Theme Park 'Fossilien Welt'. A 3D reconstruction in the paleontological era had already been made in 2014 for an area of approximately 50 km^2 of the terrain of the Korneuburg basin. The new objective is to narrate the history of the ocean and its inhabitants using animated 360° scenes. In this context, an application was created to let the visitors download it on their smartphones and have a complete guide of the site, optionally seen in an immersive 3D vision with the use of a lightweight VR headset specially developed for the purpose.

Another relevant example is *Ullastret, 250 B.C. a virtual reconstruction of an Iron Age Town.* The project consist in a digital storytelling experience based on a 3D model of the iron-age archaeological site of Ullastret in Empordà, Catalonia, developed in two different applications: an audio-visual for an on-site immersive room that consent exploring the ancient towns through a six-minute overview of the digital model and an adaptation for VR headsets, that generates more complete and realistic experiences because of the total immersion in the virtual space.

The project *Piranesi "oltre il reale"* forms part of this scenario and differs from it at the same time. It is a sensorial labyrinth that can be walked in an immersive experience using a VR headset. It is a path, a process of discovering through contrasts from a dark infinite vertical labyrinth to a bright infinite horizontal one. The visitor can feel the sensations of the Piranesian space during a process of discovering of the spaces, while his awareness of it grows. It was born from a meticulous research on Piranesi's *Carceri d'invenzione* and the consecutive definition of a Piranesian space and of its characteristics. Thanks to this experience the visitors of the Museum of Representation can relate the physical exhibit of the series of *Carceri d'invenzione* and the virtual space, thought as an expansion of the museum where itself becomes the object of the exhibition.

4 Method

The project aims to allow the museum visitors to experience a Piranesian space from the inside. It enables them to better understand the sensations that Piranesi wanted to express with the series of etching *Carceri d'invenzione.*

As seen in the previous paragraph it is necessary to change the concept of museum with a new one, that integrate new technologies in the traditional visit of the museum, in order to be more appreciated by today's audience and encourage the interaction. So, creating a virtual walkthrough inside a Piranesian space was the best way to reach the purpose of the project. An immersive experience in this sensorial labyrinth would have let the visitor to feel the emotions that designed that space.

The project has passed through many phases:

- An extensive research on Giovanni Battista Piranesi and his work, focusing on Carceri d'invenzione.
- A process of extracting concepts and defining what a piranesian space is, ending with a cloud of words that contains them all.
- The ideation of a physic space from these concepts and its continuous transformation.
- The built of this space in 3D in both physical and digital way, creating a maquette and a 3D digital model on Rhinoceros 6.
- Creation of the virtual experience by importing the Rhino model into Unreal Engine 4, using the tool Unreal Datasmith available for Unreal Studio: setting the light and materials effects; creating a virtual walkthrough; setting interactions via blueprints tools.
- Creation of the virtual reality workstation inside of the Museum of Representation.
- Evaluation test of the prototype.
- Making of the final maquette to be part of the exposition at the museum (Fig. 6).

5 Case Study

5.1 Museo Della Rappresentazione

Museum of Representation (MuRa), managed by the Department of Civil Engineering and Architecture (DICAR), is one of the 21 museum that constitute the University of Catania's Museum System (SiMuA). It is in via Etnea, one of the most important streets in Catania, inside of Villa Zingali Tetto, the symbol of the advent of Art Nouveau in Catania. The villa was commissioned by the lawyer Paolo Zingali Tetto to the architect and engineer Paolo Lanzerotti and realized in 1926. The lawyer lived in there until his death, in 1969, and bequeathed it to the University of Catania. Since 1972, the University used the villa in very different ways, from seat of the Library and Documentation Center (CBD) to 'Casa della Città', after the restoration

Fig. 6 Exposition of the maquette during the visits at MuRa. *Credits* Simona Calvagna

work in 1999. Then it was closed from 2012 and, thanks to the establishment of SiMuA, finally opened as 'Museo della Rappresentazione' in 2016 (Fig. 7).

Museum of Representation aims to strengthen the research and teaching activities of DICAR in the field of architecture and preservation and protection of heritage documents owned by the Department. With this purpose were created the R^{3}D Lab and the sPrA to promote internal research and teaching experiences, as well as experiences with external institutions such as schools, Municipalities, Superintendence, private bodies and professional offices, within local cultural heritage projects.

The heritage managed by the museum mainly consists of two collections: the Piranesi's fund, that collects more than a thousand etchings of the eighteenth century, and the Fichera's fund, formed by about 1600 heliographic copies and original drawings made with different techniques. In addition to these, there are engravings and drawings by De Vico, De Rossi, Savorelli, Pannini e Camporesi.

Fig. 7 Villa Zingali Tetto, Catania, Giardino d'inverno. *Credits* Giuseppe Tuttobene

The permanent exposition also includes the work of the Laboratories: a digital exposition of 3D reconstruction, photogrammetric and laser scanner surveys realized during educational or research activities that can eventually be seen with a VR headset by scanning the QR code on the smartphones; and the exposition of the more representative *maquettes* (physical models) realized by the students during lessons or workshop.

5.2 *Piranesi and Piranesian Spaces*

To better understand the project, it is necessary to have some knowledge of Giovanni Battista Piranesi and the spaces he created in his works.

Piranesi was an engraver, architect and architectural theorist that lived his life mostly in Rome. He was born in Venice in 1720 and, after a period practicing in engineering and architecture office, decided to move to Rome to learn the art of engraving. He was incredibly fascinated by the Ancient Rome, he spent days and nights observing every single detail and drawing them. The most part of the work of his life have Rome as protagonist. This love he felt for the ruins, for the magnificence of another era that was going to be forgotten in some time, for the immense buildings he imagined from the remains started a process of creation of spaces.

This process gave birth to the series of engravings *Carceri d'invenzione*. They took shape from this interior world, like a dream. The first version, published in 1745, was evanescent, defined by soft lines that ended in clouds of white smoke. It was not already defined; it was an idea and he wanted to visualize it. The second version, reworked and published in 1760, showed the spaces in all their parts. The lines were darker, and it gave materiality to the drawing. Towers, bridges, arches, stairs, ropes and torture instruments and again in an infinite loop. Infinite spaces, out of scale, accentuated contrasts, a more defined symbology are the characteristics of the *Carceri d'invenzione*.

This is for sure the most known and interiorized work he did in his life, as well as the several etchings of the series *Vedute di Roma*. He dedicated his life to represent the city of Rome as it was in eighteenth century, working constantly until his death in 1778. His son, Francesco, continued the work of his father and preserved the copper plates, that are now located in the National Institute of Graphics in Rome.

From the analysis of the *Carceri d'invenzione* and of the opinions of the critics expressed over time, we can define what a Piranesian space is (Fig. 8). It is characterized by exaggerated proportions of space, illusory perspective effects, tendency to infinity, uncertainties on the boundaries between external and internal environments, pronounced effects of contrast light-shadow, elements that break the view, the labyrinthine spaces.

It is an immense sequence of spaces expressing a complex vision of the world, affected by the dreams dealing with an ancient period and the reality of the world of the eighteenth century. It is a labyrinth for the mind, where the observer lose himself while trying to find a connection, a way out that does not exist. It represents the real

Fig. 8 Word cloud of the Piranesian space concepts. *Credits* Federica Grasso

world where we are constricted, the society that rapidly evolves inside of spaces that no longer belong to the people. The sensation is of being prisoners of this spaces, prisoners of life. The only way to stop roaming is to become aware that place is just a representation of the real world and we cannot escape from it, we have just to learn how to live in it.

The labyrinth then is the most appropriate instrument to represent the Piranesian space. It is a very remote origins' symbol, appeared for the first time in the Ancient Greek. Its meaning changed over time: from a path to a center that was difficult for his deep meaning and not for his complexity in shape, to a very intricate path where the difficulty stayed in finding the way out because of dead ends or multiple pathways. In English this difference is clearly expressed in the term's 'labyrinth', for the first meaning, and 'maze', for the second.

The interpretation of the labyrinth has always been twofold, from a physical place of bewilderment and perdition to a mental place in where to face fears to find oneself. With Piranesi this duplicity is combined in the anguishing representation of the places within the *Carceri d'invenzione*, maybe the first example of three-dimensional mazes.

The man, within the labyrinth, is called to understand his role because only through the attainment of such knowledge he can live there without being one of those sketchy figures wandering in the background. Perhaps it is losing orientation that he can find the exit.

The labyrinth becomes a journey: a psychological journey into oneself and a physical journey into life. The intricate pathways, defined by the difficulties of life, lead to the center of the consciousness, where the person must 'kill the beast' in order to rebirth and exit. The traveler finds himself in an infinite cycle of spaces in succession, a nightmare with boundless, infinite dimensions. Time slows down, until it stops, and gives us the view of a moment. A representation so dynamic, however, to make it appear real.

The human figure appears small in comparison to the spaces, sometimes so small as to look like an ant. The man loses his way inside the prisons, roams, lose himself and then tries to find himself again. The discovery of oneself takes place in the acceptance of disorder and of the contradiction of things, in the criticism of the ordinary and in the discovery of the hidden.

6 The Project: The Virtual Sensorial Labyrinth

Piranesi "oltre il reale" starts from the concepts described in the previous paragraph to create a new virtual space. It is a virtual prison of mind, a place where the visitors can live the Piranesian space, instead of just observing it. VR allows to approach differently to the exposition: walk inside of it, in this case.

The Piranesian space can be considered an infinite place, inside of which the person loses himself not only in the physical sense, but above all in the mental sense. The discovery only takes place after the acceptance of the disorder and the contradiction of the 'goods', it lays in the criticism of the ordinary and in the discovery of the hidden. It is with the immersion in a sensory experience that allows such self-discovery. It's necessary to be prisoners of this infinite space and to become aware of it to finally live without the constant need to find a way out.

It is a sensorial labyrinth, created from feelings to let the people feel them through the space. It is a path of self-knowledge where different types of contrast are experienced: from the emptiness to the excess of forms, from the sizes out of scale to the smallness of the man, from the psychological upheaval to the awareness of reality as it is, from the obvious to the hidden, from chaos to rule, from darkness to light (Fig. 9).

The experience, introduced into the new exhibition at the noble floor of MuRa, begins when the visit at the Piranesi's section ends. The last room of Piranesi's exhibition is dedicated to multimedia contents, the VR experience is situated over there. This experience is configured as a virtual extension of the museum dedicated exclusively to the famous series of engravings. The museum itself becomes the exhibit. By wearing the viewer, the connecting door between the museum's multimedia room and the virtual exhibition it's conceptually passed through. The visitor falls into the Piranesian space and the course to discovery and self-knowledge begins.

The space was created from the breakdown of a roman labyrinth (Fig. 10). The breakdown generated two contrasting levels: the lower one tends to vertical infinity, it is an internal space, underground, composed by horizontal plans connected by flights of stairs accompanied by walls. The visitor is forced down the labyrinth, where he can feel the disorientation, the anxiety, the vertigo and experiment darkness and silence, until the last platform. That is the connection place with the upper level. This level is exterior, bright and chaotic, composed by an infinite expanse of parallelepipeds that follow the original path of the roman labyrinth. The own concept of labyrinth had a breakdown in here, becoming a maze with infinite ways of passing through. No more walls, just white concrete blocks of different heights (Figs. 11 and 12).

Fig. 9 Sketches made during the making process of the project. *Credits* Sketches by Federica Grasso

The connections between the two levels is the light. From the upper level light goes down passing through skylights; the rays go down, defining the spaces by the light rays on the darkness.

The experience passes through these spaces, gradually exploring and revealing their characteristic and the contrasts: from out of scale to human proportions, from darkness to light, from silence to chaos, until arriving to the final block (Figs. 13 and 14).

This block, conceived as a third level, embodies the characteristics of both upper and lower level in a last element, where awareness is reached. It is at the exit of the original labyrinth path, that correspond exactly with the starting point of the experience and marks the end of the psychological cycle. It is a space like those seen before: on the outside it seems a big white concrete block, on the inside it

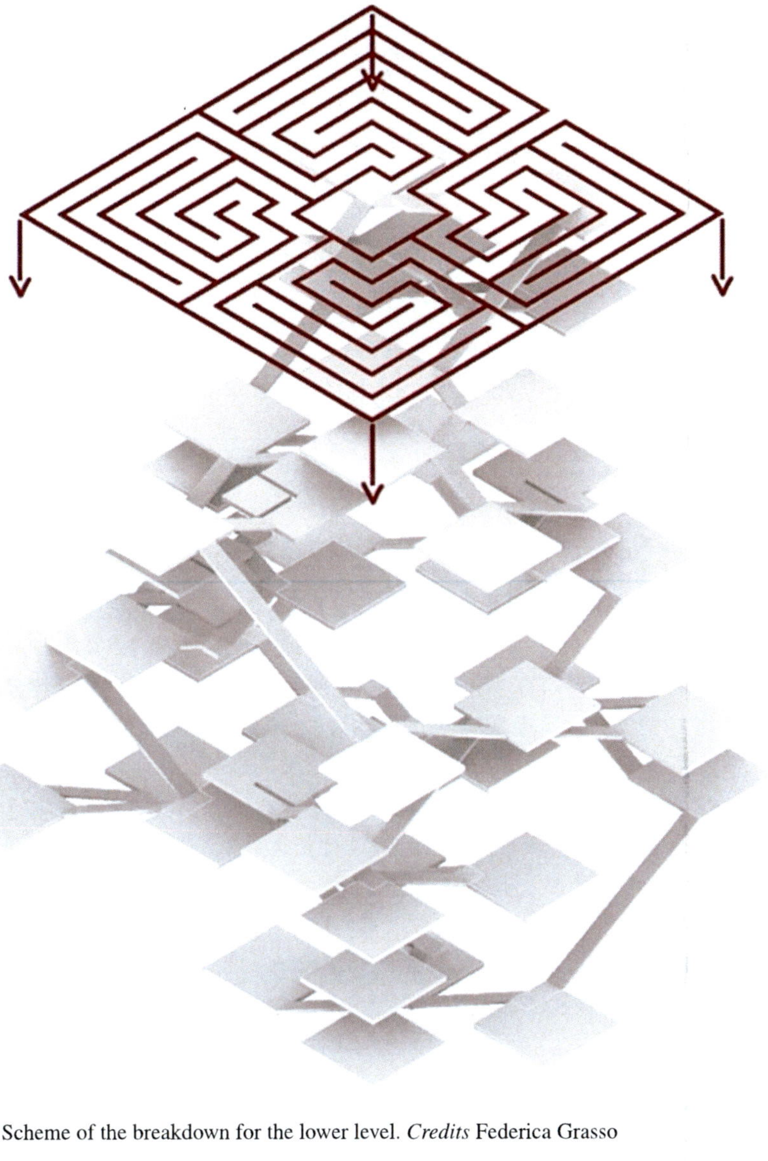

Fig. 10 Scheme of the breakdown for the lower level. *Credits* Federica Grasso

takes up the platform-stairs composition but, this time, human-sized. The feelings of bewilderment and anxiety disappear, now the visitor is aware of the spaces he crossed. The sequence of stairs leads directly to the last platform, to the roof height, opening the view to the newly crossed maze. This is the moment in which the visitor realizes that everything he has gone through is nothing but a reflection of the Piranesi's *Carceri d'invenzione* and of its feelings and concepts. It happens thanks to the anamorphosis of the Plate XIV from the second version of *Carceri* applied to the floor (Fig. 15).

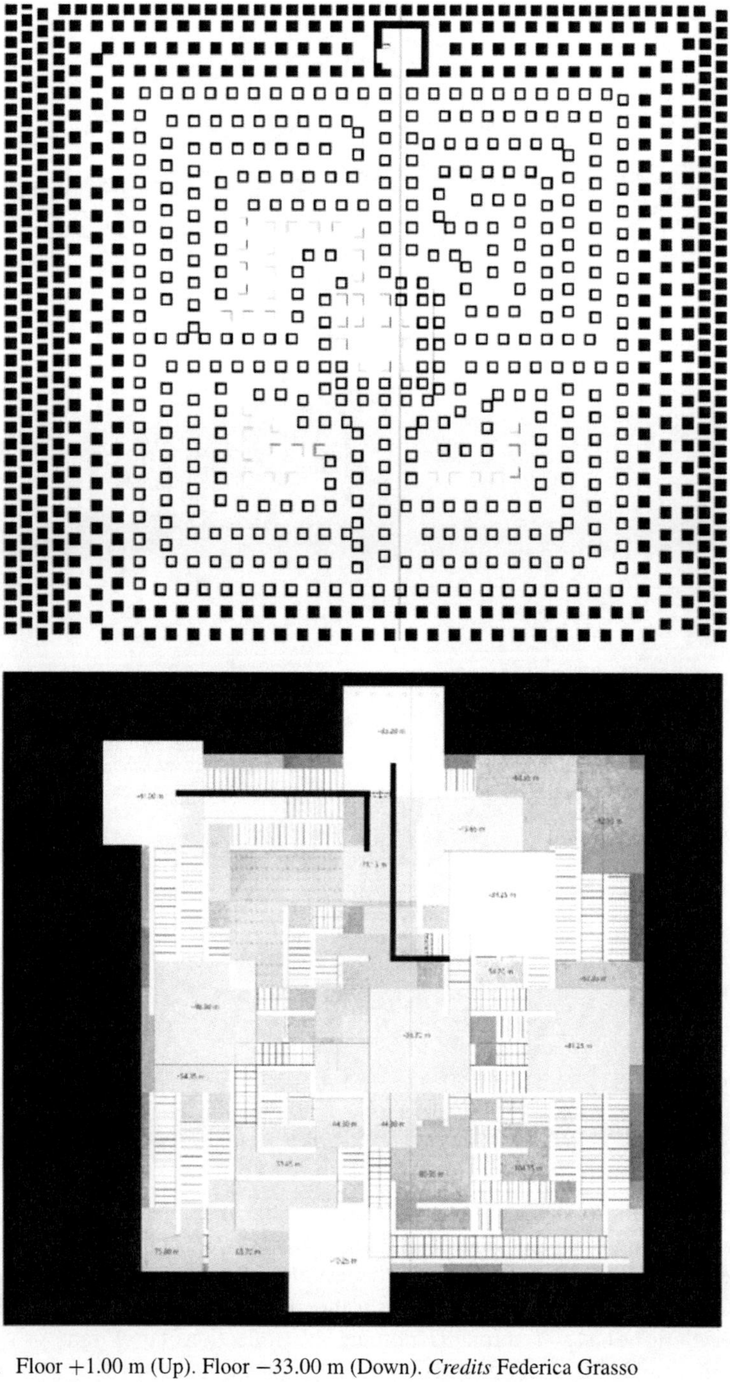

Fig. 11 Floor +1.00 m (Up). Floor −33.00 m (Down). *Credits* Federica Grasso

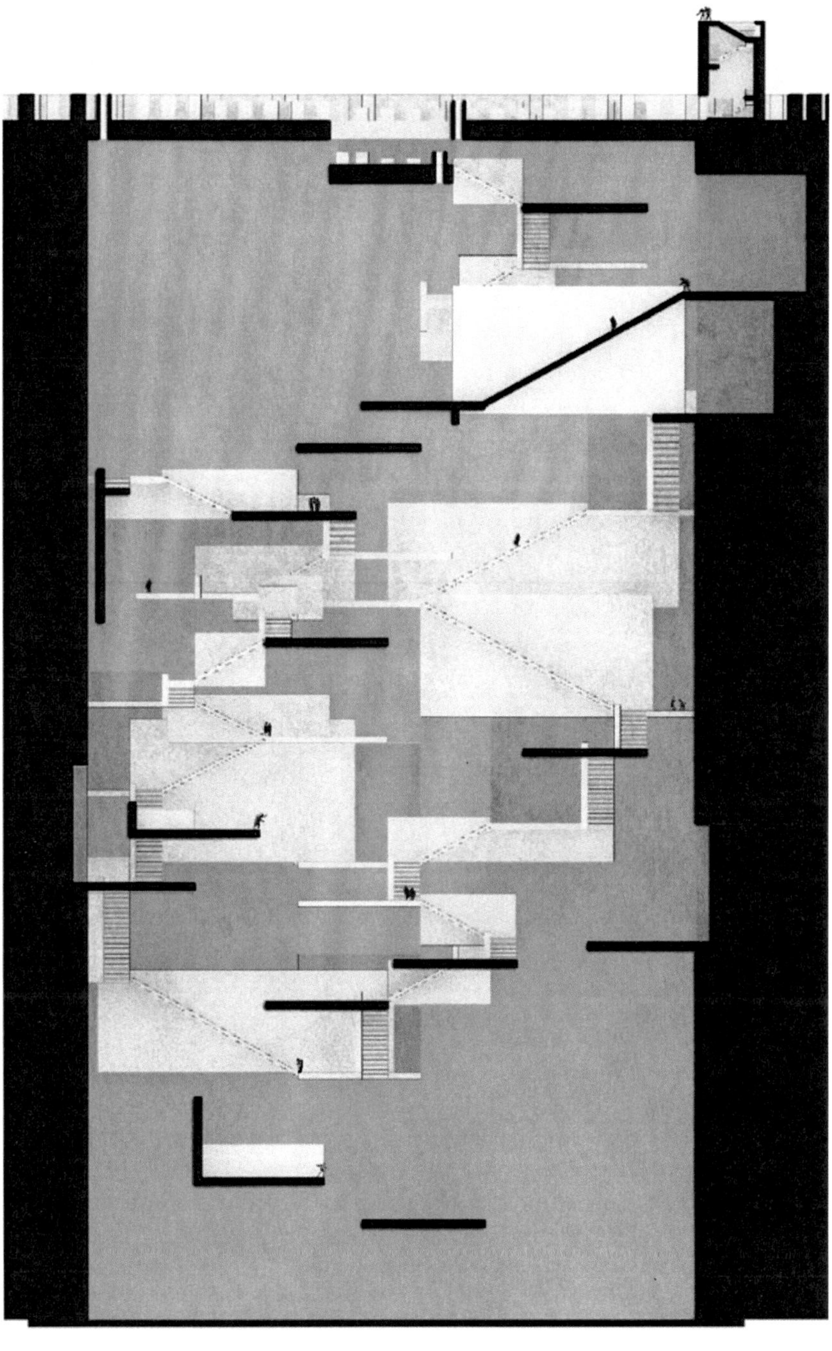

Fig. 12 Vertical section of the Piranesian space. *Credits* Federica Grasso

Fig. 13 Renders of the Piranesian space Piranesi "oltre il reale". A virtual sensorial labyrinth inspired by Giambattista Piranesi's carceri d'invenzione. *Credits* Render by Federica Grasso

The project was realized using Rhinoceros for the 3D model and Unreal Engine 4 for the optimization of light effects and materials and to make the experience explorable in VR. A physical model was also created to help the creation of the spaces in a 1:200 scale. The model was then transformed to the final definitive by printing in 3D the upper level and by using a CNC milling machine to cut the valchromat and the metal for the lower level.

The prototype has been tested many times and the results of the evaluation tests are very positive, evidencing the effectiveness of the project on the understanding of the Piranesian space and the feelings expressed in it.

Fig. 14 Render of the Piranesian space Piranesi "oltre il reale". A virtual sensorial labyrinth inspired by Giambattista Piranesi's carceri d'invenzione. *Credits* Render by Federica Grasso

7 Conclusions

In this chapter we talked about Virtual Reality in museums, sensoriality and the process by which to create a virtual sensorial labyrinth.

The project responds positively to the identified needs: thanks to the VR experience of the project *Piranesi "oltre il reale"*, the series of etchings can be better appreciated. By trying the experience, the visitor finds itself inside of a space created from the feelings of the *Carceri d'invenzione*, inside of a Piranesian space. This means that the visitor can experiment it from the inside, he can be immersed in that space, feel it and understanding it, instead of just looking and observing the plates of the series from the outside.

Fig. 15 Anamorphosis. view from the top of the final block. *Credits* Render by Federica Grasso

New virtual technologies, specifically Virtual Reality, are certainly a valuable tool for increasing the interest of visitors in the works of museums. The understanding of the works is also improved by virtual experiences, as the evaluation tests shows.

The project, currently inserted in a temporary exhibition on labyrinths at MuRa, will be integrated in the permanent one with the new set-up of the museum. The workstation will be located near the room dedicated to the series of etchings *Carceri d'invenzione*, in order to have the complete experience by observing and analyzing the etchings and then by living a Piranesian space in an immersive experience with the VR headset.

In the future the VR experience will be improved by creating a Serious Game that will allow people to interact with info panels or with drawn characters of the series and to play little games to help the discovery process.

Piranesi "oltre il reale" it's strictly connected to the *Carceri d'invenzione* series and it's not linked to one place. It is a virtual extension of the Piranesi exhibition; it has been created for MuRa but could be moved to others Piranesi's expositions if requested in order to give a different perception of the spaces represented in the series of etchings.

References

Accatino S Cárceles de invención. El Arte de Mirar, La Panera, p 11

Angelino M-I (2003) I labirinti: il dramma del percorso. Dr Virtualis (2):119–127

Artner-Rauch U, Klein M, Sovis W Immersive AR/VR museum guide. https://www.vi-mm.eu/project/immersive-vr-ar-museum-guide-for-the-miocene-site-of-stetten-austria/

Baccheschi E I 'capricci' del Piranesi. Speciale Roma, pp 46–51

Becattini M (2013) La Lumiere Mysterieuse: Ciò che ha visto Piranesi e ciò che oggi vediamo noi. In: Ananke 70, new series, pp 62–75

Berti G Labirinti della memoria. Lab 2.0 magazine 14, 12–2015/1-2016, pp 16–19

Bevilacqua M (2015) Piranesi, Giovanni Battista, Treccani, Dizionario Biografico degli italiani 84
Bevilacqua M (2008) Piranesi. Taccuini di Modena, Roma, Artemide
Bianchini R (2016) Quando i musei diventarono virtuali. Inexhibit. 10/06/2016
Bianconi GL (2012) Elogio storico del cavaliere Giovanni Battista Piranesi, celebre antiquario ed incisore di Roma. supplement to Horti Heperidum, II:2
Carobene A Realtà virtuale—approfondimento. in Treccani, Enciclopedia online
Chiuppesi M (2016) Musei virtuali e inclusione sociale. In: Rivista Trimestrale di Scienza dell'Amministrazione, Pisa, 1/2016
Economou M, Sosnowska E, Young H EMOTIVE hunterian museum digital storytelling about the antonine wall. In: Pujol L (ed) https://www.vi-mm.eu/project/emotive-project-the-exploration-of-egalitarianism-digital-education-kit/
Ejzenštejn SM (1980) Piranesi o la fluidità delle forme. translated by Maria Fabris, in Tafuri, Manfredo, La sfera e il labirinto: avanguardie e architettura da Piranesi agli anni '70, Torino, G. Einaudi, 1980, appendix
Freina L, Ott M A literature review on immersive virtual reality in education: state of the art and perspectives. Institute for Educational Technology, CNR, Genova
Galluzzi P (2010) Museo virtuale in XXI Secolo, Treccani, Enciclopedia italiana
Holl S (2004) Parallax. Architettura e percezione, Milano, Postmedia, translated by Antonella Bergamin, original edition 2000
Pérez GA (2018) Surrealism and architectural atmosphere. Arch Des 88(2):25–29. Special number: celebrating the marvellous: surrealism in architecture, marzo/aprile 2018
Realtà virtuale. Treccani Encicl Line
Roncato S (2007) Piranesi and the infinite prisons. Spat Vis
Ronco M (2011) Semantica ed estetica del labirinto. Parol Line
Tafuri M (1980) La sfera e il labirinto: avanguardie e architettura da Piranesi agli anni '70, Torino, G. Einaudi, pp 34–110
Universitat Pompeu Fabra, Ullastret, 250 b.c. a virtual reconstruction of an iron age town. https://www.vi-mm.eu/project/ullastret-250-b-c-a-virtual-reconstruction-of-an-iron-age-town/
Wilton-Ely J (1997) Piranesi, 2nd edn. Electa, Milano, pp 101–112
Yourcenar M (2017) La mente nera di Piranesi, 2nd edn. Pagine d'Arte, Tesserete, Svizzera
Zumthor P (2006) Atmosfere. Ambienti architettonici. Le cose che ci circondano, 3rd edn. 2015, translated by Emilia S original edition. Electa, Milano
Zumthor P (1998) Pensare architettura, 3nd edn. 2015, translated by Maddalena Disch and Francesco Dal Co., original edition

From Survey to Parametric Models: HBIM Systems for Enrichment of Cultural Heritage Management

Sandro Parrinello and Anna Dell'Amico

Abstract The paper summarizes some research experiences conducted by the DAda-LAB laboratory, of the Department of Civil Engineering and Architecture of the University of Pavia, on the theme of the digitization of large monumental complexes and, in particular, on the development of databases and 3D models for Cultural Heritage management. The connection between the digital survey and HBIM models is addressed by evaluating the most appropriate strategies for the production of parametric models by analyzing the methodologies aimed at the transformation of data from continuous systems to discrete models. From the digital acquisition phases to the development of complex information systems, drawing becomes the tool that allows to decompose and recompose the complexity of reality, organizing databases that offer opportunities for the development of management systems. The models in this way make it possible to explicate information according to specific parameters for the benefit of a more effective representation for the knowledge, management, and documentation of the Cultural Heritage.

Keywords HBIM · 3D model · Survey · Database · Heritage

1 The Representation in Digital Models[1]

The theme of the reproducibility of the heritage of historical cities keeps the debate open on the problems of structuring and sharing a typology of language made of signs and symbols that can be codified and recognized through the construction of generalizable formal schemes.

Through the drawing discipline, the construction of a graphic language is developed in which the complexity of reality is reduced in favor of the communication

S. Parrinello · A. Dell'Amico (✉)
DICAr, University of Pavia, Pavia, Italy
e-mail: anna.dellamico@unipv.it

S. Parrinello
e-mail: sandro.parrinello@unipv.it

C. Bolognesi and D. Villa (eds.), *From Building Information Modelling to Mixed Reality*, Springer Tracts in Civil Engineering,
https://doi.org/10.1007/978-3-030-49278-6_6

Fig. 1 Three-dimensional modeling environments: the import phase of two-dimensional files in a three-dimensional environment

of selective information. The drawing makes explicit those descriptive needs that animate the communication style. Those representative tensions focus the gesture of representing in its being a method of communication and finally, the transmission of information (Fig. 1).

The narrative system to which a specific drawing or model refers to is, therefore, part of a language which, in its compositional grammar, can be interpreted based on the concept of drawing from and for, evaluating the reality that one wants to represent and the purposes of the representation of it, through a reflection, which in some cases is implicit, also based on metric data acquisition methodologies (Cervellini 2016).

The Information can be transferred instantly anywhere and have the ability to reach anyone provided with a network connection; physical space is no longer considered enough and needs its digital extension in which objects are enriched with finite information that can become something else. The information transposed from real space into virtual space creates new digital and physical configurations that actively influence each other (Ratti 2017).

Cities, places where information are concentrated, are represented through their digital alter ego, which is nothing more than a shareable spatial expansion on which a greater number of apps and technologies are developed, that go to reproduce information and services based on different data characterization criteria (user/use). In this image extension that moves a media reformulation of the identity of the places and, more generally, a reformulation of the identity of the representation, the documentation of the historical heritage deals with the theme of the digital reproducibility of spaces through their conversion into communication platforms.

If we consider each city, its places and monuments as unique and unreproducible works, the problem of identifying the references through formal schemes is proposed, through the use of generalizable and repeatable graphic symbols of reference on which to be able to outline representation guidelines that can reproduce and communicate different types of information (Colarossi and Lange 1996): from those aimed at the visualization of the object to those of in-depth analysis of a descriptive technical nature relating to the metric data, parameters and attributes that allow digital models to be organized and interrogated based on the attributes that define them (Osello 2015).

If the models discover their function of simplification of reality, it is in the immersive fruition and in the ways of interacting with the digital data, in the interoperability and in the possibility of adapting the information to the different communication strategies that the image extension takes shape.

In the renewal of the informative message, the model once again faces the tensions that connote its formal identity, modifying the morphological metric reliability and the qualities that increase its verisimilitude and correspondence to reality, as a function of parameters more characterizing other aspects not strictly connected to the shape. If on the one hand the tendency is to make the digital model correspond as closely as possible to the real object, the synthesis of complexity is oriented on criteria of approximation of form and limitation of imperfection in favor of standardization of the formal buildings components and easier computation and interaction with the model element.

They are the criteria of use and communication that define the ways of constructing the shape to find in the graphics and therefore in the definition of a graphic language of the model lemma useful to make it easier to understand concepts that can also be complex.

In the history of infographics, an example of how a simplified type of language can be effective is given by Oliver Byrne[1] as early as 1847 in the six books "The elements of Euclid in color" having to deal with mathematical theorems through the use of a more simplified graphic language. Which is not only a simplification of the complexity of the analytical language but also become an innovative form of reinterpretation of that same complexity which allows reforming other aspects as well.

Through the definition of a graphic language, our architectural model represents and re-presents itself, as in the "The Temple of time", a map drawn at the beginning of the nineteenth century by Emma Wilard,[2] as a system that contains quantified and sized information, made discrete, limited in some way through graphic language (Fig. 2).

The design of a model implies the reading of a complexity that is filtered by an idea, an idealization of the real that is designed and therefore interpreted according to a semantic reading to become part of a form that is identified into the parametric model (Fig. 3).

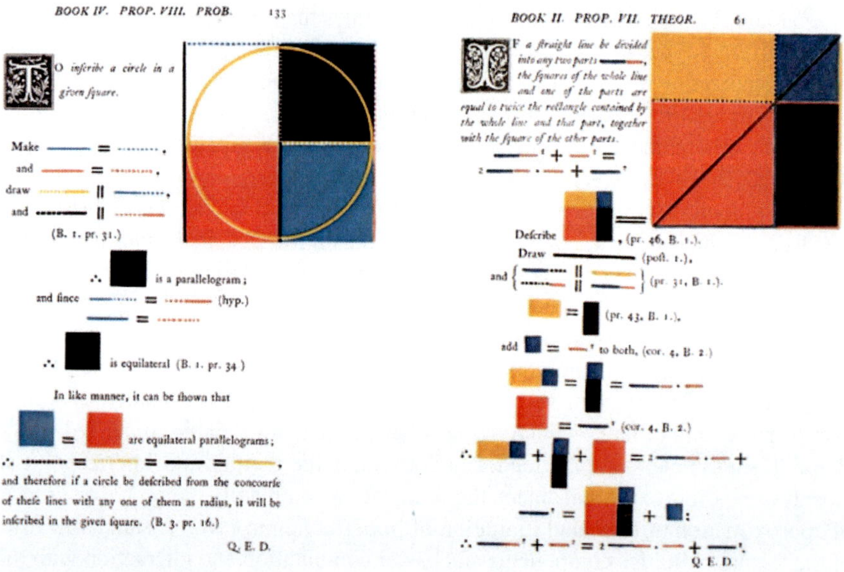

Fig. 2 Some pages taken from one of Oliver Byrne's books "The elements of Euclid in color"

Fig. 3 The infographic map "The temple of time", Emma Willard, early nineteenth century

2 The Discretization of the Continuum[1]

The information, qualified by a different nature in terms of content and specificity of the data, finds meaning through the use of specific languages, codes in which the data take a specific and contextualized value. This context guarantees a possibility of classification based on criteria and descriptors which contribute to building indexed abacus.

In the semantic reading of a built architectural continuum, this abacus can match, in a first simplified form of the information problem, to families of architectural elements, that are abacus proper to the model. These categories of elements require ordering attributes and qualities, also going to make explicit otherwise unexpressed particularities to accumulate the informative value of the model by adding to the more geometric-spatial qualities that use its shape. That condition of analysis of shape using a decomposition in space and the character that C. Norberg-Schulz describes well in his Genius Loci and that seemed to have found so much confirmation in the production of documented strategies based on the elaboration of digital information maps (Bertocci and Parrinello 2007), here seems to renew its sense in the development of parametric databases that are not only associated with a geometric component but that are an integral part, an inseparable whole in which the reading of the structure of the place through levels or components finds its correspondence in a universe of models organized by classes and families (Fig. 4).

In this modulation of the model shape and the redevelopment of the graphic structure of the architectural drawing, a tension is revived that concerns the reliability of the drawing and the need to guarantee that quality standard that in recent decades, through digital technologies, has characterized the research for the survey and the architectural representation.

In this perspective, reality-based models are drawn starting from the data acquired through the architectural survey and it becomes useful to evaluate which iteration systems involve the method of construction of the model to define the most appropriate methodological protocols to guarantee reliability.

It is useful to remember that a parametric model does not have as its main purpose the physical description of the asset, but intends to represent a much more complex

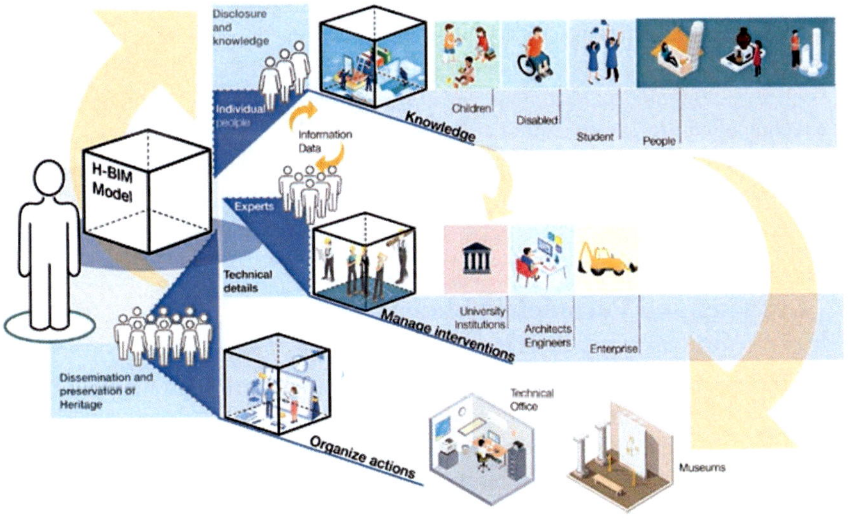

Fig. 4 Diagram of the figures involved in the management processes of the H-BIM model

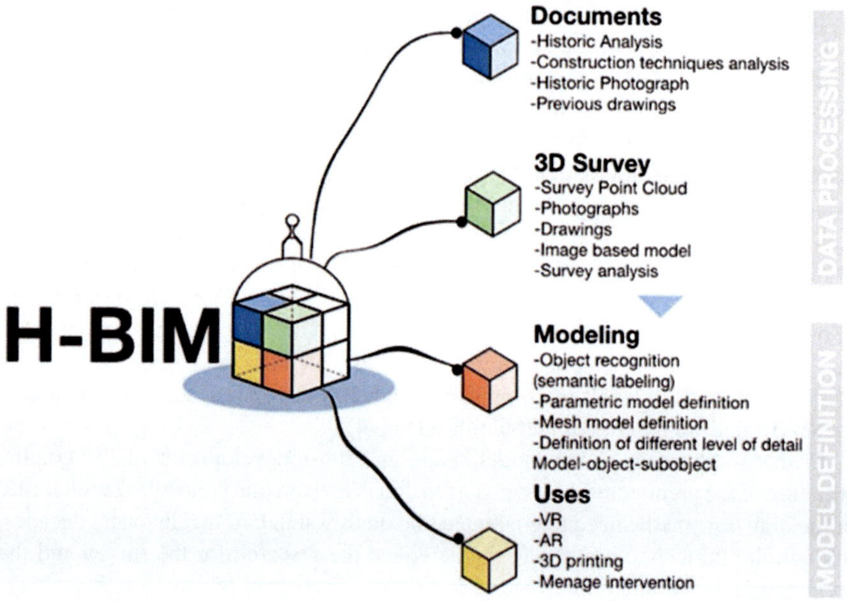

Fig. 5 Diagram of the elements that define the connotations of the three-dimensional model

system by configuring itself as a sharing information tool. The model is the shared tool that allows more people to work and cooperate, collaborating through the setting of a common language used through the same platform that finds its image in the three-dimensional model (Osello and Ugliotti 2017).

The morphometric data, connected and inseparable from its structure, qualifies instruments where different actors collaborate concerning the purposes of the process. Within a single model, which can be configured as an HBIM system, data useful for the enhancement, use, management, and control are inserted. The parametric model, of a technical nature, thus becomes a shared tool that also performs functions that are not purely technical, of the didactic, exhibition and promotional activity of cultural systems (Fig. 5).

3 Drawing and Parametric Model[2]

The software that gives the possibility to model and link information elements to the models for the management of buildings based on the BIM concept, although limited in some functions since they are specifically built for the ex-Novo design of buildings, do not exclude the possibility of construction, through a thoughtful project of information model modeling aimed at documenting the built heritage.

The evolution of the static model (NURBS) into a parametric dynamic model benefits from the advantages of information management guaranteed by the BIM protocol.

The BIM application to the Heritage sphere implies an action of documentation of material and immaterial elements of historical structures and environments. The examples provided by the literature for cataloging and collecting data for the translation of these into an information model include buildings belonging to categories of highly protected buildings with more complete intervention projects and careful management of the life cycle.

Research on these topic by Murphy and Dore proposed and analyzed six elements of the HBIM process: heritage documentation standards, data collection techniques, 3D modeling concepts, as-built BIM and procedural modeling (Jordan-Palomar et al. 2018).

Of particular importance is the study on the development of standards for the sector conducted by Carleton Immersive Media development of standards for its use in traditional applications (Dore and Murphy 2017).

Speaking about cultural heritage, it is necessary to define a methodological protocol different from the approach used for modeling the design of buildings. The modeling actions require a definition of the data acquisition phases and the levels of knowledge and in-depth analysis to define a correct modeling procedure. H-BIM is focused on a protocol aimed at translating the information of the metric data, as resulting from the laser scanner survey operations, into an archetype model. The model geometries can be the result of the data editing by the comparison with the software family libraries, modifying its attributes where possible and necessary until a geometric synthesis of the forms is achieved or through in-place modeling operation of the individual objects drawn ad hoc for the different case studies (Li et al. 2019).

At the basis of the modeling processes, it is good to declare the purposes of the model that will be built.

The first operation necessary for the definition of the model is the decomposition of the architectural continuum based on the analysis and cataloging of the elements to identify the appropriate modeling strategies for each of them. The elements are identified according to the first distinction between structural and decorative elements. Among these, after having defined the different categories of the model, the replicable elements that are those categories belonging to the model for which once the basic parameters have been defined are identified: it is then possible to repeat them within the model. These types of elements will be constructed through the structuring of a specific project family. For those types of elements that cannot be replicated within the model, the choice of an in-place modeling for the representation of the data as a generic model is more effective.

Once the different elements have been analyzed, and an indexed abacus of the objects that will make up the model is structured, it is possible to proceed with setting the project file.

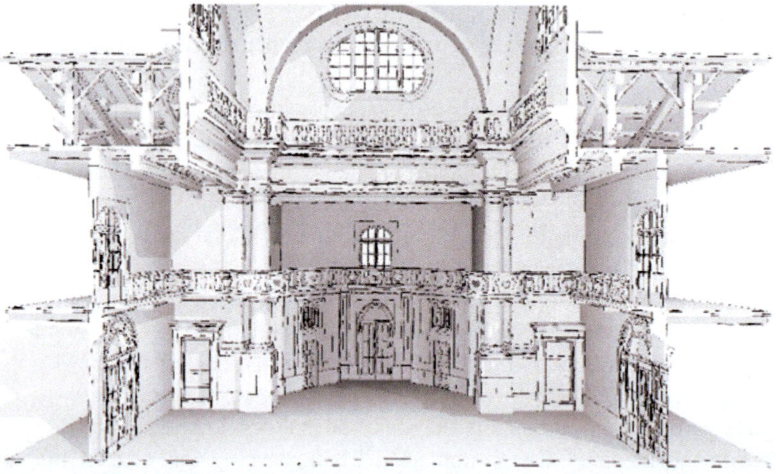

Fig. 6 An example of a HBIM model, in the cross section you can appreciate the complexity and diversification of the elements present in the model

The traditional protocol provides for the import of two-dimensional files (.dwg), like floor plans and sections, as useful traces for setting the reference levels on which to then anchor the models of the different element type families.

However, if we are working with existing buildings this requires the obligatory step, an additional phase that foresees the operations of drawing the data of the digital survey in a 2D environment. This time-consuming phase is not always effective for continuous monitoring of the reliability of the metric data modeled in comparison with the acquired data (Fig. 6).

The use of the two-dimensional drawing limits the geometric information to the data already processed which brings with it a first evaluation and interpretation of the real data which will inevitably add to the second interpretation of the data by those who will model and then draw for the second time the same data in three dimensions. The paradox and the question, that arises spontaneously, is: why, if the acquisition from laser scanner technology provides a three-dimensional type of data, for the elaboration of the models a passage must be then made to drawing in two dimensions and then return during the phases of the elaboration of the model into a three-dimensional environment? For this reason, methodological protocols are developed for the reduction of modeling times trying to cancel, as much as possible, the two-dimensional drawing phase of the elements. Through specific plugins[3], it is possible to view the point cloud data within the software Revit. The use of plugins for viewing the point cloud has the advantage of reducing the waiting times due to the possible import, export and conversion of the scan file data into a point cloud format compatible with Revit, found that the default software prefers dialogue with the proprietary program Autodesk ReCap. This type of visualization is useful for not overloading the file in terms of bytes (Figs. 7, 8 and 9).

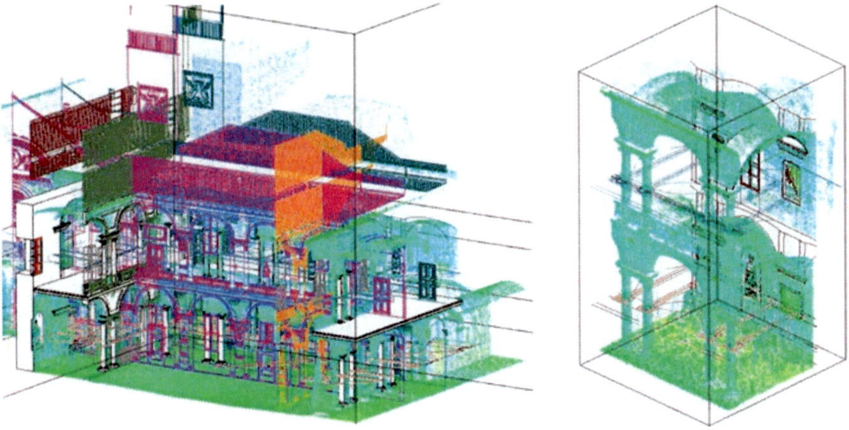

Fig. 7 Comparison between two-dimensional drawings dwg, point cloud data and 3D model imported in the same modeling environment

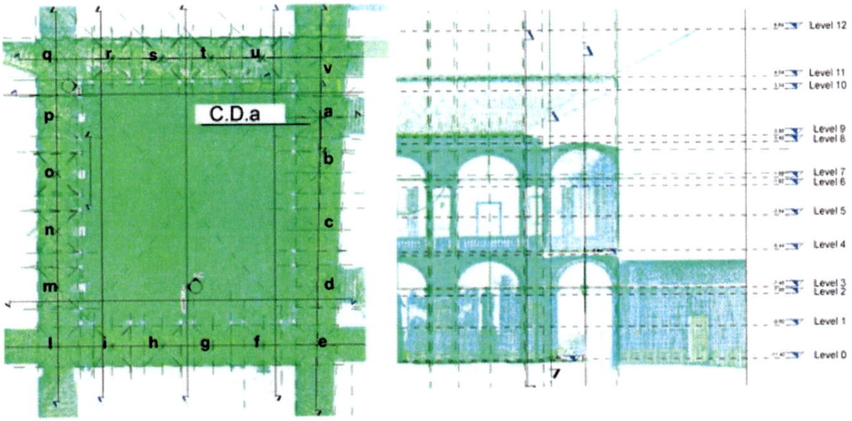

Fig. 8 The setting of the reference planes based on the point cloud data

Once the point cloud has been visualized through CloudWorx, it is possible to set the different reference levels that go to structure an ordered and indexed grid which will be the skeleton of the model, useful during the subsequent modeling phases. This display method offers the advantage of having continuous control over the correspondence and adherence between the three-dimensional data of the metric measurement and the 3D drawing of the elements.

For elements such as walls, floors, roofs, stairs, it is often possible to adapt families pre-set in the software by changing their dimensional and stratigraphic information by editing the model parameters and directly modifying the drawing lines through control points. For elements with more complex geometries such as vaults, columns, windows, balustrades or decorative elements such as statues, plaques, frames, it is

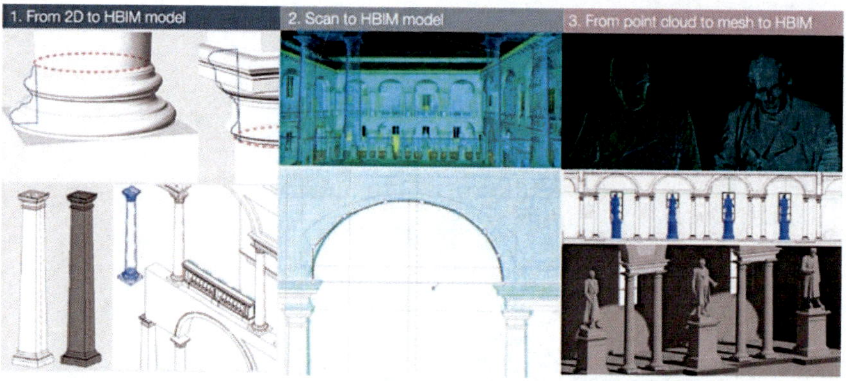

Fig. 9 The three different modeling approaches: using 2D CAD drawings, using the morphometric information directly using point cloud 3D databases, importing complex geometries after the point cloud 3D meshing

necessary to envisage a different model strategy. For these types of elements, it is recommended to evaluate whether to create a family for the element to be included in the model or whether to model it directly in the project environment as a generic model, then associating it with a specific category of elements. The example of the modeling experimentation conducted on the courtyards of the headquarters of the Central University of Pavia is here reported.[4]

The modeling phases of the experimentation can be divided on the basis of the three different approaches used for the representation of the different geometric complexities of the object of study. The first one based on the import of the point cloud and the Scan to BIM modeling through the creation of the model elements using families of systems, a second one dedicated to the enrichment of the descriptive detail of the model through the loading of families of elements modeled ex novo on the basis of two-dimensional drawings, and finally a third phase of mesh modeling and import of the complex geometries of the statues present in the museum.

In the first modeling phase it was possible define through the plug-in connection the basic structure of the model by defining walls, roofs, floors, stairs using families of systems present by default within the modeling program (Fig. 10).

These systems have been adapted and modified in their basic parameters through a direct control of the design of their geometry trying, by setting control points, to match the geometry of the model as much as possible to the morphometrics data given by the point cloud (Morandotti et al. 2019).

Once the basic wrapping of the model "box" was defined, the individual elements were modeled.

For those components for which it was not possible to use the predefined models of the elements, we opted for the creation of specific model families, for those elements that appear to be replicable within the environment, dividing them neatly according to their typological characters (windows, columns, doors, balustrades) or through

Fig. 10 View of the three-dimensional model of the courtyard of the statues (Unipv) during construction

the creation of a generic geometric model drawn in-place using the point cloud data as a guide.

In the modeling phases of the two vaulted open gallery, we can find the application of the different modeling methodologies. Given the irregularity of the elements also due to the different inclinations of the wall facings which, if observed in the planimetry, are not parallel to the wall of the colonnade, for the three-dimensional representation it was necessary to build a model escamotage by identifying a geometric solid, in this case an ellipsoid, which was as close as possible to the profile of the section in key of the vault and then go to subtract from it the excess solid material through subtraction solids configured on the reference geometric construction lines

In this case, given the presence of a double order of twin columns, the individual columns were analyzed: the data found were all different from each other imperfections due to labor and wear and tear dictated by the passage of time. For this reason, the representation of the element by type was preferred without going into the modeling detail of the individual column models with their own differences. Three model families have been created for this purpose, divided according to the macro types of column identified during the census of the elements (Morandotti et al. 2019).

For the representation of the model families, given that the software does not allow the same possibility of viewing the point cloud in the family editor, it was necessary to import the 2D profile drawing of the column elements, the result of the processing of the acquired data by laser scanner through which it was possible to set the different categories of model, whose construction took place through extrusion and revolution operations for the generation of model solids.

The third modeling methodology was used for the representation of the elements with complex lines and geometries such as the statues and elements present inside

Fig. 11 3D model views

the courtyard. It was preferred to represent them using the import of mesh models generated by the processing through specific processing software. Mesh surfaces of the point cloud (Geomagic DX and MeshLab). The omission of these elements, which characterize the environment, giving it a specific identity and recognisability between the different rooms of the building would have deprived the model of essential narrative information related to the place. Once converted with a file extension compatible with the software (.dxf) these mesh models are recognized by Revit as static models and therefore cannot be modified, it is however possible to create a specific family for each element that allows you to attach specific information to the model purely descriptive character through the structuring of model parameters (Fig. 11).

4 From General to Particular, the Model as an Information Container[2]

The research promoted in the DAda-LAB experimental laboratory of the University of Pavia, deals with the development of parametric models, as containers of information, which can interface and be suitable for the different levels of analysis and technical studies of the data. The definition of a hierarchy of virtual environments understood as containers of specific information useful for the management and development of heritage and territory management implies the possibility of defining analysis models that can be developed using the criteria of architectural representation and that can be implemented with the contribution of numerous researchers.

If on the one hand the quality and quantity of information that can be collected on the architectural object has increased and is increasing with technological development, on the other this trend entails a heterogeneity of the formats used for the documentation of the heritage, thus generating an accumulation of files with different data extensions that complicate communication and exchange between databases and professionals (Calvano 2019). New storage needs and different protocols for data management are required. The structuring of a platform as a large digital container that can collect different types of information, even if not necessarily compatible with each other, is still a utopian project and leads to continuous collisions with limitations problems in terms of the specific weight of the files. In the common imagination when we think of the digital world, thought associates it directly with an undefined space, limitless, with infinite possibilities; in the real context of daily use these limits still exist and during the operational phases we come across different problems due to data overload that often makes the platforms not very fluid and unmanageable or to machines in the possession of the typical user who are unable to compute a large amount of data.

The experiences aimed at managing and disseminating information relating to cultural heritage through information systems are numerous and continue their digital evolution. To be effective and useful for the management and dissemination of information, it is necessary to design the information model, in which not only are the essential guidelines for the construction of the information container model outlined but where the same information must be managed based on the diversity of contents. The information must be conveyed in a coordinated way for the management of the various contents and purposes of the operators involved in the conservation, protection, restoration, and dissemination of the heritage, who besides being users of the model become its main developers (Pavan et al. 2017).

To be able to manage information, so that it can be accessible, a management system coordinated with the tools and systems used during the life cycle of the model asset is required. Therefore, the main characteristics that a system must have are interoperability, accessibility, standardization (Agustín and Quintilla 2019).

The use of digital models for the documentation of historical heritage requires, therefore, in addition to the definition of modeling processes, also the drafting of a data sharing protocol that orders the methods of information exchange between the various disciplines and technologies. Currently, the web is the place that offers the possibility of sharing and accessing the content.

The representation of the model through different levels of detail (LoD), allows to provide a fluid visualization and efficient data access. The information modeling systems can integrate various types of information and documentation, as well as offering to different users the possibility of interconnection with the web, they allow you to actively participate during the model structuring phases (Lo Turco 2015). The involvement of various active figures in the enrichment of information requires the need for structuring between the different users a shared methodological protocol document such as a manual created specifically for the individual project that is useful to define the guidelines for the implementation of the model.

The conservation and management of the built heritage is a complex process that requires analysis at different scales, a multilevel that arises from the collaboration between different specialists in the sector.

The ordinary information that can be read by the HBIM model can include (Pocobelli et al. 2018):

- Geometric data, from the metric survey;
- Definition of schedules on architectural elements;

- Definition of building materials. Which can be expressed graphically by the use of crosshatch, textures or by simple annotation labels;
- Surface degradations that can be described through the use of UNI glossaries and graphically expressed through the use of crosshatch and labels;
- Construction phases.

To these are added a series of metadata that can be duly linked to the model to deepen the information level:

- Archive photos;
- Texts/archival research documents;
- Data on environmental analyzes;
- Cost data;
- External links that allow connection with secondary models for the deepening of geometric data or use links to websites and different databases on the cloud (Pocobelli et al. 2018).

A type of modeling according to the definition of different information levels through the orderly organization of the elements in a digital guise has the intention of extending the life of historic buildings by shaping a new reality of our historical—cultural memory into a digital memory aimed at creating new accessible spaces (Parrinello 2019) (Fig. 12).

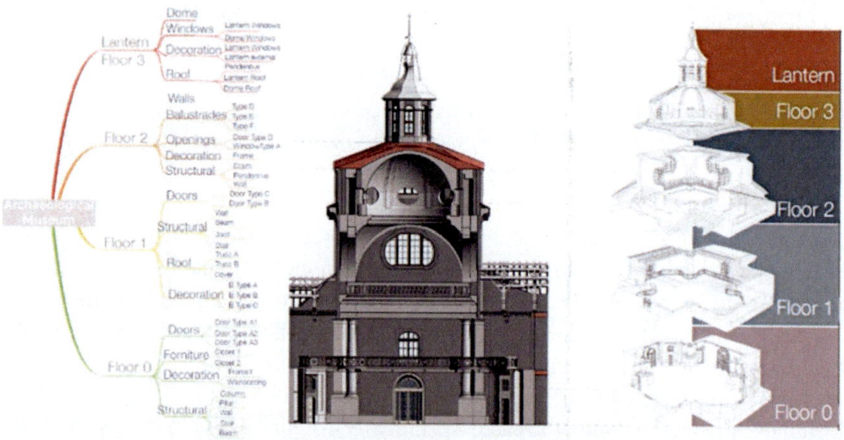

Fig. 12 The breakdown of the elements that characterize the H-BIM model, based on the category type

5 From Particular to General Through the Use of Virtual Environments[2]

Museums and historical monuments are adapting their visit program to the ongoing social transformation. By observing the behavior of society, their content and the type of cultural offer available to citizens are changing, implementing real data with the necessary digital extension. Around the H-BIM digital model, a series of possibilities orbit according to the different fruition systems that can range from didactic, exhibition, dissemination, promotional and management activities. The programs that allow the visualization and multimedia navigation of BIM models and contents are in continuous development and updating. Two different VR visualization platforms available on the market at the time of the Autodesk Live and Twin Motion BIM experiment were tested within the research conducted. Both platforms support the use of models from PC-desktop and through the use of VR viewers (Oculus Rift). The Twin Motion platform is a free plugin that allows the direct connection between Revit and the Twin Motion editor (Unreal Engine rendering engine) with maximum compatibility and interoperability between elements. It is possible to insert the model within different scenarios, changing the environmental and lighting conditions of the scene. The model is displayed complete with the materials set on the Revit platform, however, the plugin gives the possibility to change the materials also directly using the presets set by the software material library. Lacks the possibility of deepening the information of the data, the objects are selectable and are listed by categories but it is not possible to view any additional information. The Revit Live platform works as an Autodesk cloud service. The model is sent by the Revit Live software plugin link to the cloud platform which processes the conversion of the model and sends it back to the user in a format ready for sharing the scenario with the extension.lvmd. The application is more limited in the construction of the scenario, compared to the previous one, not due to the possibility of changing the background of the landscape and the materials but maintains the characterization of the scene and material settle down on Revit. Precisely because of the compatibility guaranteed by the Autodesk manufacturer, in this case it is possible to view the information and parameters of the individual objects through information sheets structured by default and not editable. During viewing, it is possible to measure the distances from the position within the user's virtual environment with three-dimensional objects. Unlike the Twin Motion plugin, the Live is a service that Autodesk doesn't offers for a fee, the trial version limits the use of 10 uploads in 30 days. Being a sector that of virtualization and compatibility between Revit and fruition platforms, everything is still in the development phase, plug-ins and developers change quickly with system updates and present new opportunities. The same Revit Live used for this experimentation in a few months will be obsolete will no longer support the updated versions of the Revit software already the 2020 version does not have this option in its interface, because it has been replaced by new plug-in possibilities such as Unity Reflects born from the collaboration between Unity and Revit release scheduled for December 2019. This fact indicates that the technology is under development and the need for management protocols to be continuously monitored and kept under control also based on the possibilities provided by the different developers, adapting in some way at the time of development of management software (Figs. 13 and 14).

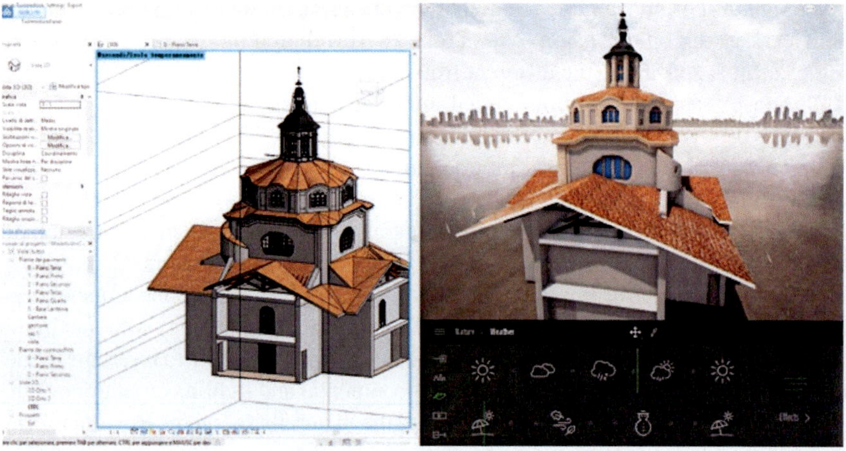

Fig. 13 The connection of the model to the twinmotion viewing platform

Fig. 14 Comparison between the two plugins (twinmotion and revit live) for the interactive visualization of the H-BIM model

6 Conclusions[1]

Synthesizing reality through the development of a model implies defining in turn models based on a process of logical abstraction of forms. The digital transposition of heritage, for sharing and communication with the global society, through communication channels based on open access and social access and sharing systems, continuously searches for new possibilities of configurations and data visualization.

Advanced digital models can produce a digitalization action of global architecture, frame objects in spatial modules, structure the analyzed characteristics and transform the output of the tools into optimized 3D databases, responding to the dynamic and standard requirements of European regulations.

The data collected on site during survey activities, through the different tools, explain an overabundance of information which, if not processed and downloaded, loses their communicative function. The set of data and information, in order to become a communication tool for different users, must find the right means of explanation. Drawings, symbols, geometries, images, sounds and multimedia elements have their peculiarities and equal information dignity if justified by a specific purpose (Parrinello et al. 2019).

This is because from the analysis of the general data to the in-depth analysis, an attempt is made to understand how the parametric model is, on the one hand, interfaced with the database, on the other, it can become an enhancement tool to encourage the use of multimedia content. The model generated within the HBIM management platform can be exported and used in secondary fruition platforms for various purposes.

Speaking of historical heritage through the computer literacy of the diffused heritage it is possible to redirect tourist flows, giving visibility to those monuments or to those itineraries that are less known and frequented by the people masses, as well as allowing access at any time through the making available of a widespread digital museum, which does not want to be a substitute for the real visit experience, but rather a tool for deepening available to the visitor-widespread (Maggi 2007). Faced with an ever-changing panorama in which the tools of representation and use of the elements vary and are diversified, research is directed towards a process of the evaluation process on the potential of the various HBIM scenarios for the enhancement of historic environments. Following the directives of the new building and territorial management regulations of international policies such as the funding programs allocated by the European Commission H2020 and the various regulations that national and international policies are adopting, the research aims at defining through experimentation on different case studies of protocols for the management of the historical heritage built through HBIM modeling. Today, the use of diversified and unrelated protocols between the various players in the sector are generating a high quantity of digital products which, however, are disconnected from each other, generating a dispersion of information as well as generating ineffective working practices that hinder interoperability also causing economic losses, the goal of the future development is to enclose in a single modeling protocol different possibilities of use and programming through the use of collaborative systems that allow a better sharing of information in heritage construction projects.

Notes

[1]Oliver Byrne (c. 1810–c. 1880) was a writer and civil engineer of Irish origin. In addition to a treatise on euclidean geometry, the book is a new didactic expedient in which Byrne uses graphic symbols and primary colors to express more complex concepts. The author states that this is a didactic device that distinguishes its edition from all the others: "Euclid's Elements in which symbols and colored schemes replace the letters, to the benefit of those who approach this subject".

[2]Emma Willard (1787–1870) American activist in 1814 founded the first high school reserved for female study.

[3]There are different possibilities of dialogue between the point cloud and the Revit software depending on the software used to manage the point clouds. In particular, the laboratory is experimenting with the use of the CloudWork plugin proprietary of the Leica Cyclone software. The plugin allows you to view the cloud by exactly replicating the settings set on the Cyclone software.

[4]The research is part of a project commissioned by the technical office of the University of Pavia that requires the digital documentation of the entire complex of buildings in the historic headquarters of the central University. Significant activities began in December 2018 which was attended by qualified staff from the DAda-LAB Laboratory together with students from the University of Pavia. The research on the modeling of the complex was concentrated on the courtyard of the Statues and the rooms of the Sala della Crociera of the ancient St. Matteo Hospital, now home to the university archaeological museum on this thematic is undergoing a graduation thesis by the student Fu Hangjun.

References

Agustín L, Quintilla M (2019) Virtual reconstruction in BIM technology and digital inventories of heritage. Int Arch Photogramm Remote Sens Spatial Inf Sci XLII-2/W15:25–31

Bertocci S, Parrinello S (2007) Rilievo e Piano di Gestione per il Centro storico di Montepulciano. In: Clini P, Lancioni N, Quattrini R (eds) atti del convegno EARCOM 07 Sistemi Informativi per l'Architettura. Alinea, Firenze, pp 108–113

Calvano M (2019) Disegno digitale esplicito. Rappresentazioni responsive dell'architettura e della città. Aracnee, Roma

Cervellini F (2016) Il disegno come luogo del progetto. Well's Srl, Roma

Colarossi P, Lange J (eds) (1996) Tutte le isole di Pietra. Gangemi, Roma

Dore C, Murphy M (2017) Current state of the art historic building information modelling, Int Arch Photogramm Remote Sens Spatial Inf Sci XLII-2/W5:185–192. https://doi.org/10.5194/isprs-archives-XLII-2-W5-185-2017

Jordan-Palomar I, Tzortzopoulos P, García-Valldecabres J, Pellicer E (2018) Protocol to manage heritage-building interventions using heritage building information modelling (HBIM). Sustain MDPI Open Access J 10(4):1–19

Li D, Wang X, Bai C, and Wu C (2019) Discussion on the problem of regularized reconstruction in HBIM. Int Arch Photogramm Remote Sens Spatial Inf Sci XLII-2/W15:657–662

Lo Turco M (2015) Il Bim e la rappresentazione infografica nel processo edilizio. System Graphic, Roma

Maggi M (2007) Musei alla frontiera. Continuità, divergenza, evoluzione nei territori della cultura, Jaca Book, Milano

Morandotti M et al. (2019) L'università di Pavia, i cortili e gli ambienti monumentali. Un progetto di documentazione digitale e lo sviluppo di sistemi di gestione per la manutenzione programmata. In: Conte A, Guida A (eds) Conference Patrimonio in divenire Conoscere, Valorizzare, Abitare, Reuso VII international conference. Gangemi Editore, Matera. 23–26 Ottobre 2019

Osello A (ed) (2015) Building information modeling geographic information system augmented reality per il facility management. Dario Flaccovio, Palermo

Osello A, Ugliotti FM (eds) (2017) BIM: verso il catasto del futuro. Conoscere, digitalizzare, condividere. Il caso studio della Città di Torino. Gangemi Editore, Roma

Parrinello S (2019) Digital and real: parallel processes of documentation and project. In: Parrinello S (ed) Digital & documentation. Databases and models for the enhancement of heritage. Pavia University Press, Pavia

Parrinello S, Picchio F, De Marco R, Dell'Amico A (2019) Documenting the cultural heritage routes. The creation of informative models of historical russian churches in upper kama region. Int Arch Photogramm Remote Sens Spatial Inf Sci XLII-2/W15

Pavan A, Mirarchi C, Giani M (2017) BIM: Metodi e strumenti. Progettare, costruire e gestire nell'era digitale. Tecniche nuove, Milan

Pocobelli DP, Boehm J, Bryan P et al (2018) BIM for heritage science: a review. Herit Sci 6:30

Ratti C (2017) La città di domani: come le reti stiano cambiando il futuro urbano. Einaudi, Torino

"BIM Survey". Critical Reflections on the Built Heritage's Survey

Tommaso Empler, Adriana Caldarone, and Maria Laura Rossi

Abstract Legislative Decree 50/2016 and the subsequent Ministerial Decree MIT 560/2017 create an obligation, in Italy, for the digitization of building contracts. This is an adjustment, in terms of internationalization, regarding the adoption of BIM processes in the building field. For new design these processes are widely shared both in terms of interoperability and transparency between various professional figures involved, but above all thanks to the containment of costs of planning, building and management of the work in its life cycle. Applying the same processes to existing historical buildings, of which our country is most composed, does not find the same consensus: fundamentally due to the need to use standardized components that do not adapt to peculiarities of a historical architecture, where an obvious geometric reading is not always possible. Therefore, a whole series of problems arise concerning the documentation, communication and representation of the historical building as well as important implications in the field of surveying, very often mentioned in tender procedure as a prerequisite for the construction of a BIM model. This scenario opens up questions of an epistemological nature which are partly addressed in this document. A survey path, in the direction of "BIM Survey", is carried out by testing the potential in new forms of modeling/information through visual algorithms (VPL) on some parts of the Municipality of Accumoli, following the earthquake that struck central Italy in 2016.

Keywords BIM survey · VPL · Knowledge process

T. Empler (✉) · A. Caldarone · M. L. Rossi
Sapienza University of Rome, Rome, Italy
e-mail: tommaso.empler@uniroma1.it

A. Caldarone
e-mail: adriana.caldarone@uniroma1.it

M. L. Rossi
e-mail: marialaura.rossi@uniroma1.it

C. Bolognesi and D. Villa (eds.), *From Building Information Modelling to Mixed Reality*, Springer Tracts in Civil Engineering,
https://doi.org/10.1007/978-3-030-49278-6_7

1 Introduction

Legislative Decree 50/2016 and Ministerial Decree MIT 560/2017 defines methods and timing for the progressive introduction, by tender stations, granting administrators and economic operators, of mandatory methods and tools for digital modeling in BIM (Building Information Modeling), in tender for building and infrastructure, in design, construction and management of works in their life cycle.

Parametric modeling processes, linked to the use of libraries of standardized objects, such as BIM, now widely used in the definition of new designs, are little known in the field of existing buildings and, above all, of historical buildings, which constitute a large part of the architectural heritage of a country like Italy. Among others, are important the consequences related to the use of HBIM processes (Heritage or Historical), as regards the knowledge and documentation of the work (historical, material, technological information, anthropic transformations or degradation occurred over time, etc.), but, above all, in the field of surveying, since it is not always possible to recognize the geometric nature of a historic building and associate it, in a single model or in a single digital environment, with all the information necessary to describe it.

Furthermore, an extremely heterogeneous and multidisciplinary field, such as that of cultural heritage, requires an increase in interoperability, considering the inclusion of professional figures not necessarily foreseen in the field of new design (superintendencies, restorers, etc.). Despite perplexities and questions, which change case by case and to which we still do not know how to give objective and shared answers, the obligatory nature of the decree imposes in many tenders, as a support for modeling, a digital survey.

More and more often it happens that, in tender procedures of public and/or ministerial offices, a so-called "BIM survey" is put out to tender, whose definition is however unknown both to the offices that draw up the announcement, and to those who intend to participate in it, leading to evaluation and transparency difficulties. A question arises spontaneously: does "BIM survey" exist and can we talk about it?

Both operations (integrated digital survey and BIM modeling of the built) constitute models that give back the meanings of architecture: the first, following a survey project based on the recognition of salient features of architecture, returning a digital discrete numerical model (point cloud); the second is a parametric and informed continuous model, resulting from a critical synthesis of the input data and of the "semantic" recognition of components and their relationships.

At present, BIM modeling applications acquire point clouds directly (they take on the status of blocks, objects that cannot be modified once imported into the software), which must be "vectorized" with parametric library objects made available by the same software (Bolognesi 2018; Carta et al. 2018; Catuogno and Di Luggo 2016; Di Luggo et al. 2017).

This process involves a "reduction" in terms of adherence and geometric accuracy with respect to the survey input data since the standard architectural component, even if properly programmed, will never be able to virtually match the complex "shape" of real objects/buildings. Consider, for example, a wall with a bulging, whose geometric shape becomes primary in a path of restoration or consolidation. This swelling must then be associated with other information, by the number of layers that make up the wall and by internal gaps, which are not homogeneous but localized, to the causes that led to it: infiltration, structural transformation through its own or induced static stress, etc. Most of the digital models built in BIM environment do not foresee at all that the standardized component "wall" can be different from a perfectly planar surface (single layer or multilayer). At the most, it is planned to provide the client thematic documents that highlight the problem.

But in the context of a "BIM Survey" request, is it possible to bypass the dichotomy between representative and real form so easily? There are procedures for surveying and restitution that allow us to "inform" a digital model (by its ideal nature) with heterogeneous data and sources, representative of both the design intentions underlying the work and the state of fact and possible alteration in which does it consistently apply to BIM purposes?

2 Background and Critical Reflection

The cause of the absence of detailed answers to the above questions, is to be found in a lack of a unique technical/scientific lexicon, which doesn't leave space to interpretations, before the legislation takes it into the incomplete state in which it is found. Often, we find in the indications of the bids some specific wordings, actually used in the academic field and above all in the scientific disciplinary sector of the representation (SSD ICAR17). This is not comforting, because the use of this terminology in the application field is not reflected and has a purely evocative character.

For example, the term "BIM Survey" probably indicates the use of 3D Shape Acquisition tools and methods (integrated digital survey techniques: terrestrial laser scanner, aerial photogrammetry, image-based modeling or structure from motion, etc.) such to obtain a discrete point-like numerical model as a basis for future modeling. The fact that BIM applications are able to import and read information contained in point clouds does not automatically determine a paradigm change of survey techniques, established and consolidated in unsuspected times; rather, it could be appreciated as ambitious goal of a complete integrability of data, successful for relevant digital data in BIM software, which do not arise mainly with the purpose of operating on existing building.

We can highlight, about the "Scan to BIM" formula, some definitions from the web: "It is the process that uses any laser scan to create high-precision 3D BIM models for redevelopment, renovation and renovation projects. Once the area of interest has been digitally detected, the point cloud is transformed and the BIM model is built" (orientarum.net 2017); "It is the process that captures the built environment using 3D laser scanning technology and transforms it into an integrated BIM model" (lasia.it 2018); "It is a reverse engineering process that uses advanced detection technologies, such as 3D laser scanning, to obtain point clouds and 3D meshes to be used as the basis for BIM modeling. In essence, it transfers reality-based information into BIM systems […] the BIM model thus obtained accurately reflects reality" (mum.it 2018).

What connects these sentences is a certain (erroneous) conviction that there is some automatism such that the use of point clouds, faithful to the actual state of the building, transfers the same level of detail to BIM model. These are only a few cases, but explanations of the confusion dictated by simple contradictions in terms and which, however, are the basis of the requests made explicit in the calls for tenders.

It is therefore necessary to define the objectives of the request upstream with the client in charge, in order to define methods of massive acquisition and the level of data accuracy. Next, we will define methods for returning a parametric BIM model and the related level of detail, which does not coincide with the Level of Accuracy of the survey. Furthermore, as we said, if we consider the survey phase in the broadest sense related to a 360° knowledge of a historic building, it is not possible to stop at the geometric dimensional values, but it is also necessary to include constructive, formal and cultural history aspects. The retrieval of this heterogeneous information, coming from analogical and digital sources, must then contemplate a method of integration, such that the model is able to receive it and act as a multidimensional and multidisciplinary database.

Precisely in reference to the level of accuracy of the survey, it is necessary to establish categories related to an historic building and expected return, to which the client can refer once the goals have been set. Without prejudice to the fact that the Level of Accuracy collects results of the entire relevant procedure as a summation of the accuracy of each phase, starting from acquisition, processing and recording of data and their post processing, it is possible to define three macro categories of LoA (Level of Accuracy): "low" for large-scale surveys, "medium" for surveys of single buildings or architectural aggregates, "high" for surveys, for example, of sculptural elements that require a high level of detail (Fig. 1).

Survey's Global Level of Accuracy procedure certainly influences the development of BIM model for what concerns the Level of Detail or Development (LoD), however there is no biunivocal correspondence between these factors, they do not vary

Fig. 1 Level of accuracy.'s macro categories. From left: higt, medium and low. The level of accuracy, according to which the acquisition is conducted, does not depend on the size of the object, but on the level of detail that is desired based on the objectives

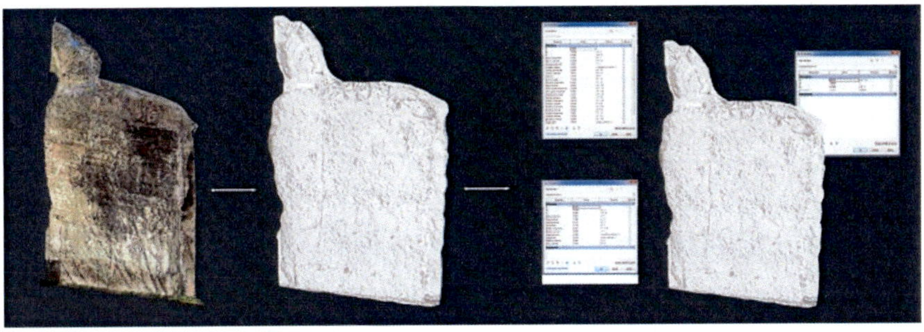

Fig. 2 Mesh information process of a claudio aqueduct's barrel-vault. The aim of the client was to calculate geometric barycenter, enter information regarding materials and compute the volumes to be restored for securing. Level of geometry is more important than level of Information

according to a direct proportionality, but according to a relation called "covariance"[1] which does not include the cause-effect relationship: it can be said that there is a systematic relationship between two variables, but not that one automatically determines the other. Furthermore, in accordance with Italian legislation (UNI 11337: 2017), the concept of LoD can be subdivided into LoG (Level of Geometry) linked to the ability to translate the information contained in the point cloud into a continuous parametric key, and LoI (Level of Information), both connected to the LoR (Level of Reliability) understood as the level of global knowledge of the process of defining a digital object (Bianchini and Nicastro 2018; Nicastro 2016). LoR is also influenced by LoA.

It is necessary to establish, on a case-by-case basis, the most suitable workflow depending on whether you want to enhance formal qualities of an object, to detriment of the level of information (structural, material, historical, etc.) that characterize it, or vice versa, to obtain a balance between the two factors, always in accordance with the target set.

To enhance geometric data, for example to preserve deformations and out of plumb to make metric calculations, you can think of working in a BIM environment directly with appropriately tessellated meshes, deduced from point clouds, skipping the parametric conversion of the object in favor of a greater formal and dimensional characterization (Santagati et al. 2018; Miatton and Parrinello 2017). Remember that in this case surfaces will be treated as objects that cannot be modified within the application, to which it is not possible associate any semantic relationship with the components of the architectural organism (Fig. 2).

Vice versa, where it is important to enhance the level of information of the object, such that the model acts as a database for documentation and data storage, it is

[1]In statistics and in probability theory, the covariance of two statistical variables (in this case the measure of uncertainty) is a number that provides how the variables change together. A positive covariance, for example, indicates that there is a high probability that there will be an increase in the second variable with the increase of the first one, or a decrease in the second variable with a decrease in the first one, although not correlated by proportional ratios.

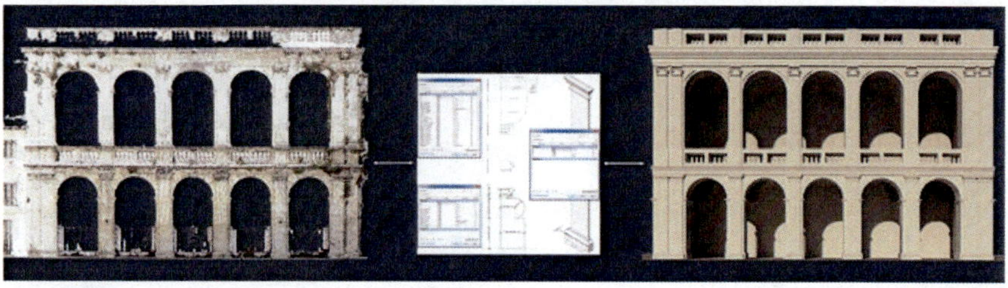

Fig. 3 Modeling process using library objects. The intent was to enter information for the documentation of the cultural heritage. Model was created by Laura Rossi

permissible to reduce adherence to relevant data and obtain simplified forms, with dimensional characteristics and proportional in principle, using standardized library architectural components (Fig. 3). In this case semantic and ontological link between components is recovered but, as we have said, all those data that are not geometrically parameterizable are neglected (Giovannini 2017).

Where we want to achieve a balance between the level of geometric definition and that related to the information of architectural component, it is possible to refer to potential inherent in new forms of parametric-generative-visual modeling, connected to the use of a lexicon universally recognized, related to computer programming, also called VPL (Visual Programming Language).

Thanks to this innovative modeling method it is possible to combine the formal and even complex aspects of architectural objects, and their semantic relationships, with a high level of information. The procedure proposed here aims to catalog and enrich digital objects of information elements even before they take on a three-dimensional configuration in a BIM environment, by sectoring and tabulating information by categories, easily associated with the models that come to be constituted (Fig. 4).

Associating continuous parametric model with the informative computational model determines a virtuous reiterative circle for which, following parametric variation of input data, it follows a controlled alteration of the global configuration through the cascade propagation of modifications. This overall model is therefore definable

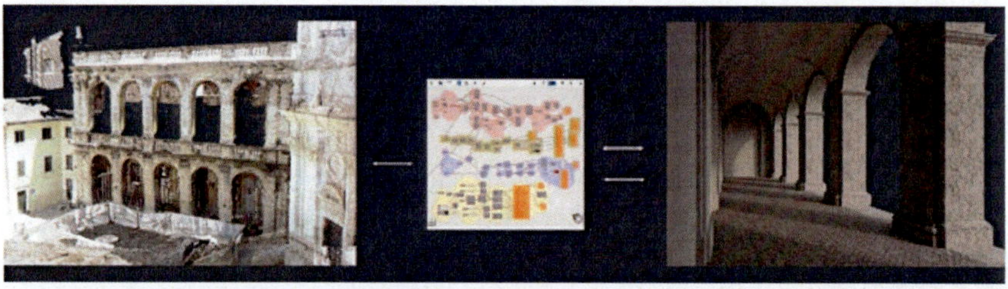

Fig. 4 Algorithmic-generative modeling process. The purpose was to establish geometric genesis of Renaissance arcades and insert useful information for restoration. The open mode allows the model to be adapted not only to the case study, but, with small variations, to other Renaissance arcades. Geometry and information have the same degrees of importance. Model by Laura Rossi

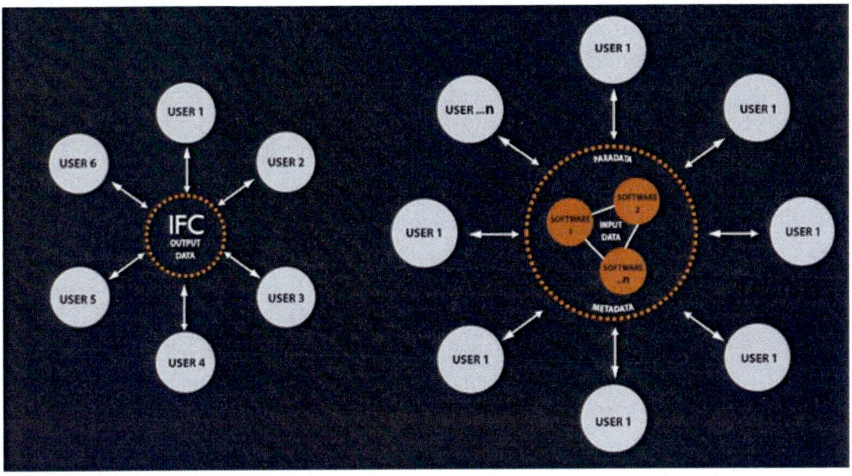

Fig. 5 Representative schemes of interoperability. From the left, a diagram showing the current status and use of the IFC format. On the right scheme of the proposed solution through the use of multiple systems and software

as "responsive" or "adaptive" (Calvano 2019): the procedure is open, always modifiable and implementable with new information entering cyclic flow, and the model is able to adapt to new conditions that generate it.

In this way interoperability, to which BIM models have always aspired, has increased considerably. Interoperability means the ability of a computer system or product to cooperate and exchange information without loss or alteration of the same information, in a reliable, certified manner and with consequent optimization of resources. In the name of this interoperability an unique open BIM format called IFC (Industries Foundation Classes) has been established: a particular data format that allows the exchange of informed geometries without loss or distortion of information, whose use has not yet given expected results. The goal will probably not be resolved with a single interchange output, which is why it is proposed here to use software and input formats, specific for each need, such as to be read simultaneously and compatible to all others by means of VPL in a BIM environment (Fig. 5).

3 Methodological Framework

An investigation path in the direction of "BIM Survey" was carried out on Accumoli and its villages following the earthquake that struck central Italy in 2016, to communicate, with organizations, professionals and more generally specialized figures, all useful information for seismic prevention processes for smaller urban centers. The procedure presented is still in experimental phase, and it is a further development of a workflow applied to reconstruction in the same context of the aforementioned earthquake, which aimed at establishing a pre-earthquake cognitive frame and a post-earthquake cognitive frame (Empler 2017).

The main is to define a procedure that is effective for issues of prevention, defining the best configuration and visualization of "informed" models, thus obtaining a "preventive digital memory" that can be implemented and multi-scaled, which sets three-dimensional model as interface of access to information.

Difficulties inherent the process are immediately recognizable: first of all the extremely heterogeneous, multidimensional and massive nature of data to be acquired must in any case be compared with a reasoned synthesis of data in a way to expand its usability and accessibility; moreover, the built heritage is presented as a reality characterized by infinite possible variations, considerable degrees of complexity that are not always attributable to a geometric recognizability of the space and to a typical standardization of BIM platforms.

Survey stands as a link between prevention and knowledge of an urban and architectural organism, and it is possible to identify it as a primary element for the cognitive path.

Detecting an urban center does not consist only in acquiring merely morphological or geometric quantitative data, but means knowing values that characterize these complexes, both through understanding of overall characteristics of the urban organism, and through description of the single elements that contribute to the formation of these characteristics. Moreover, when we talk about assessment knowledge, we refer to a whole series of surveys and studies carried out according to two paths: on one hand the archival-documentary and bibliographical; on the other hand 'direct reading' of historic buildings, analyzing materials, construction techniques, behavior of structures, possible degradation and instability phenomena, geotechnical aspects.

It is therefore necessary to develop a process of historical-critical reading of buildings that make up the urban organisms, analyzing them from a typological-building, historical-architectural, dimensional, structural and material point of view, also in relation to urban and territorial context of reference for the formation of a complete cognitive framework. Even before survey phases, one focuses on obtaining a descriptive model of the "forma urbis", as a bearer of identity features of the place and the result of historical cultural and landscape relationships. This step is fundamental for the understanding of basic building of widespread architecture, since it is the product of manual and artisanal construction activities characterized by materials, means of work used and specific ways of building in relation to the identity characteristics of the settlement.

Next step involves the phase of data acquisition, both analogical and digital, which can be divided into two macro-categories: metric/geometrical data and specialized data. Geometrical data are result of architectural surveys, specialized data are outputs of archival research and surveys diagnostics characterized by diversity of format and data Level of Accuracy. During the phases of architectural survey we focus on acquisition of data that returns information on general morphology of the buildings, their consistency, relationships between the full and the voids of urban fabrics. However, the use of 3D shape acquisition makes it possible to acquire information with greater

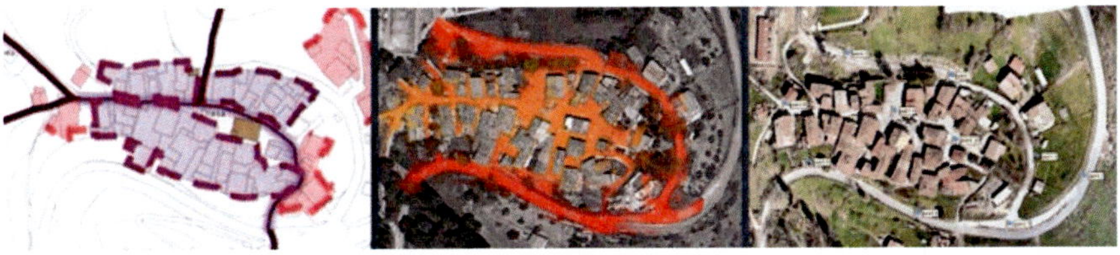

Fig. 6 Knowledge process and survey with integrated techniques. From left: Study of the urban fabric and urban fronts (studied by A. Bruschi); point cloud from survey with LS; point cloud resulting from structur from motion survey from UAV system. Restitution by Adriana Caldarone

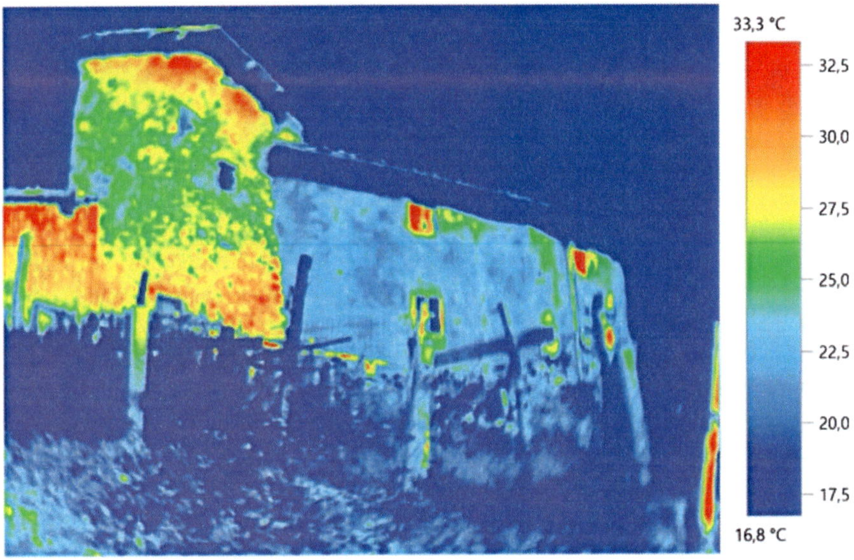

Fig. 7 Thermalgraphic image of a terracino aggregate

level of detail, especially regarding the acquisition of the shape of the buildings as a whole of its irregularities and not only in its essential lines (Fig. 6).

The collection of specialized data involves a series of non-destructive and non-invasive diagnostic investigations,[2] in particular the use of georadar and electrical tomography for characterization of subsoil and use of thermographic camera for recognition of materials and any degradation phenomena associated with them (Fig. 7). The integrated digital survey techniques allows the collection of further specialized data, especially with regard to reflectance and RGB values of surfaces,

[2]For "non-destructive" diagnostic investigations we mean all those techniques that respect the material, structural integrity of the object of study. This category also includes "non-invasive" diagnostic investigations, understood as those applied techniques in which there is no material and structural alteration, and neither does the thermodynamic state or, more generally, the chemical-physical aspects of the object be altered (Santopuoli and Seccia 2008).

which allow to obtain data regarding the masonry and their state of conservation, for less limited to what concerns the resistant and sealing layer of buildings.

In this phase it is not possible to establish the level of global accuracy of collected data, as it cannot be calculated as a simple sum of degrees of accuracy of each single gripping activity. During post-processing it will be the operator himself who will select data acquired according to the tasks set and the level of knowledge that the model will have to return. In order for have a global knowledge process and embrace all fundamental components of an urban system, all resulting information must be able to return a multi-scale model, where it is possible to trace elements in hierarchical order of decomposition and recognisability: of territorial, urban character, at aggregate level, building level and finally of single architectural element, each characterized by its own Level of Geometry and Level of Information.

Level of Accuracy, specific to survey phases in fact, as seen above, is closely correlated with the Level of Geometry and Level of Information of the informed model, and they vary with the Level of Geometry, according to a specific covariance relationship. This implies that procedure must be repetitive, that it allows, if are to be obtained data of a higher degree of accuracy, to implement and increase information by structuring hierarchical and tree-like models characterized by different LoG and LoI. The use of VPL language in restitution phase allows to obtain this advantage, in fact is elaborated a responsive model, which varies its conformation and its level of detail according to the quality of the input data in an Open mode.

In this step we focus on the organization of all the data deriving from the formation of cognitive framework (geometric, vectorial, text files, graphics, images, etc.) to be able to structure them critically. Always VPL performs two functions: on one hand it allows the association of vector and geometric data with metadata and external para-data of different formats, often transmuted and archived in alphanumeric format or as attached elements, overcoming the limits imposed by the type of information that can be associated to the digital objects in BIM environment; on the other hand it combines the qualities of a mathematical and algorithmic-generative modeler (the adaptation to complex geometries) with the semantic recognizability of a BIM modeler. The composition of architectural element and its associated attributes takes place directly in the computational environment and is imported and visualized within the parametric BIM software (Fig. 8).

In the restitution phase, starting from geometric data and from studies on the most frequent building types in the area, we focus on increasing the libraries existing in the BIM modelers through the introduction of elements and objects that return the aspects and the criteria fundamentals of the construction of the villages of Accumoli, both from a perceptive and typological point of view. The operation is descriptive of building techniques handed down over time, often with very few variations, which describes figurative characters of an architecture in relation to its geographical area and therefore can be parameterized and reusable in similar contexts.

Resulting model is a three-dimensional model in a BIM environment, containing both geometric and non-geometric information, which semantically recognizes architectural elements and binds entities together through ontological relationships appropriately identified in the formation phase of the cognitive framework (aggregates

Fig. 8 BIM model of a Terracino's urban aggregate realized through Visual Programming Language. The procedure is responsive and the model is updated and displayed in real time. Model by Adriana Caldarone

and fabrics, construction techniques and elements, architectural, materials and state of degradation). Elements that are not geometrically parameterizable, deliberately neglected (such as out of plumb, bullets, etc.), are inserted as information, or deducible from comparisons containing the deviation between simplified model and point cloud, or finally by attaching elevation map, both measurable elements, allowing thus to verify the actual value of any deviations.

To the described procedure a final phase of sharing and displaying data will be added, useful for greater accessibility and exchange of information for people involved in prevention activities.

This could be possible thanks to computational programming: annotative elements, associated with the attribute tables (metadata and para-data), described above, are connected to digital objects. In the visualization phase, these objects will become hyperlinks in the model. By selecting the object of interest, the user can connect to document containing metadata and/or para-data stored for that element.

This should allow users an exploration based on information and models that are correlated and critically structured, so that the users themselves can draw reliable evaluations and analytical results and plan interventions. This step is still in experimental phase.

Conformed model is characterized by a high level of information and a geometric development of digital objects in accordance with assessments for seismic prevention both at the urban and building level.

It can be defined as a "cognitive model" of reality that describes complex phenomena and processes that characterize urban settlements and their respective vulnerabilities.

4 Conclusions

A standard "BIM Survey" procedure has not yet been defined, although it is often referred to activities where the use of BIM is envisaged, there are tools such as Visual Programming Language (VPL) that allow processing and management of heterogeneous data, deriving from acquisitions, in function of a dynamic and continuously updatable cognitive framework that is formed by the surveyor (understood as a plurality of figures who, each for his own competences, is able to improve some levels of specific knowledge on the building subject of activities detection). VPL is connected to 3D modeling applications often associated with BIM modelers, for management of data that cannot be parameterized from the acquisition stage, such as those belonging to the built heritage.

BIM processes and their interoperability (thanks to IFC format) cannot therefore be considered as exhaustive elements, on the contrary VPL's potential is to integrate with other systems, such as semantic web and linked open data, determines an interoperability not only downstream of the process but also upstream: it does not end with the simple use of an interchange format, therefore in an output, but, on the contrary, it uses a series of input formats, and software that are able to read them and that is their interface with each other, each with its own specific peculiarity and function and which contribute to ensuring that all information, geometric and otherwise, is not lost during the process and that it is transmissible and easily communicable.

Characteristics with which BIM modelers have been devised (for new building and not for management of existing structures), require their rethinking precisely according to interoperability, the uniqueness of historical objects detected, and the possibility of inserting and managing data relating to "informed" type of cognitive framework. The methodology proposes a procedural "pipeline" where more softwares are involved (from point cloud management to BIM modeler). The effectiveness of this procedure can be obtained when all steps will be coordinated and managed in a single application environment. Clearly this produces some questions regarding sources heterogeneity and their control, regarding managing of complex databases, as well as induces us to reflect on the ability of a visual, informative and cognitive language, to communicate and be easily understood by end users who will take advantage of the model.

It is hoped that in the future, thanks to the scripting that allows the insertion of heterogeneous data all linked together, the system output can be expanded with informations deriving from smart sensors (today they are used for cultural heritage monitoring), making the even more reliable preventive framework.

Characteristics with which BIM modelers have been designed up to now (for new constructions and not for the management of built heritage), requires their rethinking based on interoperability, on uniqueness of the historical objects, and on the possibility of inserting and managing data relating to the "informed" cognitive frame.

The methodology proposes a "pipeline" in which multiple applications intervene (from point cloud management to BIM modeler). The effectiveness of this procedure can be obtained when the steps are all coordinated and managed in a single application environment, and it will thus be possible to demonstrate how it is effective in the formation of any cognitive frame applied to built heritage.

References

Bianchini C, Nicastro S (2018) La definizione del Level of Reliability: un contributo alla trasparenza dei processi di Heritage-BIM. In: Empler T, Quici F, Valenti GM (eds), Proceedings of workshop 3D modeling & BIM. Nuove Frontiere, Roma April 2018. Dei Tipografia del Genio Civile, Roma, pp 228–245

Bolognesi CM (2018) Rappresentare in 3D: il rilievo per un modello HBIM. In: Empler T, Quici F, Valenti GM (eds), Proceedings of workshop 3D modeling & BIM. Nuove Frontiere, Roma April 2018. Dei Tipografia del Genio Civile, Roma, pp 166–181

Calvano M (2019) Disegno digitale esplicito. Rappresentazioni responsive dell'architettura e della città, Aracne editrice, Roma

Carta E, Minchilli M, Tedeschi LF (2018) Dal rilievo fotogrammetrico al BIM: prospettive didattiche per la gestione e la conoscenza del costruito storico. In: Empler T, Quici F, Valenti GM Proceedings of workshop 3D modeling & BIM. Nuove Frontiere, Roma April 2018. Dei Tipografia del Genio Civile, Roma, pp 246–261

Catuogno R, Di Luggo A (2016) Dalla nuvola di punti all'HBIM. Rilievo e modellazione per la conoscenza e la gestione del patrimonio architettonico. In: Empler T, Proceedings of Workshop 3D Modeling & BIM. Applicazioni e possibili futuri sviluppi, Roma April 2016. Dei Tipografia del Genio Civile, Roma, pp 58–66

Di Luggo A, Pulcrano M, Scandurra S, Tarantino S (2017) Modellazione H-BIM e ricostruzione delle trasformazioni del costruito storico. In: Empler T, Proceedings of workshop 3D modeling & BIM. Progettazione, design, proposte per la ricostruzione, Roma April 2017. Dei Tipografia del Genio Civile, Roma, pp 178–195

Empler T (2017) Gli strumenti di rappresentazione nella procedura ARBIM. In: Di Luggo A, Giordano P, Florio R, Papa M L, Rossi A, Zerlenga O, Barba S, Campi M, Cirafici A, Territori e frontiere della rappresentazione. Atti del 39° convengo internazionale dei docenti delle discipline della rappresentazione, Napoli September 14–16 2017. Gangemi editore, Roma, pp 1559-1568

Giovannini EC (2017) VRIM workflow: semantic H-BIM object using parametric geometries. In: Empler T, Proceedings of Workshop 3D Modeling & BIM. Progettazione, Design, Proposte per la ricostruzione, Roma April 2017. Dei Tipografia del Genio Civile, Roma, pp 212–229

lasia.it (2018) Il rilievo con laser scanner ed il bim, il connubio perfetto. https://lasia.it/rilievo-laser-scanner-ed-bim-connubio-perfetto/. Accessed 20 Nov 2019

Miatton C, Parrinello S (2017) Modelli descrittivi per la gestione degli interventi di manutenzione sulla facciata della basilica di San Michele a Pavia. In: Empler T, Proceedings of workshop 3D modeling & BIM. Progettazione, design, proposte per la ricostruzione, Roma Apr 2017. Dei Tipografia del Genio Civile, Roma, pp 260–277

mum.it (2018) Scan To BIM. Un Vantaggio Competitivo per l'Analisi Digitale di Edifici e Costruzioni. https://www.mum.it/scan-to-bim/. Accessed 17 Oct 2019

Nicastro S (2016) L'applicazione del BIM come sistema informativo localizzato nel processo di conoscenza del patrimonio culturale. In: Empler T, Proceedings of workshop 3D modeling &

BIM. Applicazioni e possibili futuri sviluppi, Roma Apr 2016. Dei Tipografia del Genio Civile, Roma, pp 172–183

orientatrium.net (2017) Scan to BIM. Dal rilievo laser al modello BIM. Il passaggio alla costruzione digitale. https://www.orientatrium.net/leica/scan-to-bim.html. Accessed 12 October 2019

Santagati C, Lo turco M, Garozzo M, (2018) Eruzione e interventi antropici: modelli informativi e basi di dati per un progetto integrato di documentazione. In: Salerno R. Proceeding of 40° convengo internazionale dei docenti delle discipline della rappresentazione. Rappresentazione materiale/immateriale, Milano September 2018. Gangemi editore, Roma, pp 1561–1568

Santopuoli N, Seccia L (2008) Sviluppi delle tecniche analitiche e diagnostiche per la conservazione. Carbonara G. Trattato di Restauro Architettonico. Secondo aggiornamento. UTET, Torino, pp 190–265

BIM and Mixed Reality for the New Management of Storage Area

Daniela De Luca and Anna Osello

Abstract The fourth industrial revolution that has touched the last decades, has involved several sectors including the construction world. The digitization and automation of industrial processes has implemented the development of connection platforms that can communicate many information for different users such as Smart Glasses or immersive headset. Most of the time, innovation does not only concern the technological field, but involves the entire organizational and managerial sphere. Digitization allows new tools such as Building Information Modelling to expand its application scale, making it an excellent tool for integrating and sharing data with their own information management systems (MES). The aims of this contribution are reproducing a virtual warehouse through parametric digital modelling, to which all the management data have been associated; for example, the average stock, the rotation index, etc. Thanks to the export of the database extrapolated from its management system of the analyzed industry, it was possible to define the correct visualization of the virtual model, interrogating the real data coming from the real warehouse. Through Smart Glasses, the user of the area could allocate the products in the correct position and update in the cloud the information properties associated with the individual product and the entire department. The use of virtual platforms for the visualization and the sharing of the data, facilitate the optimization of the industrial processes.

Keywords Mixed reality · Industrial processes · Building information modelling · Digital twin · Industry 4.0

D. De Luca (✉) · A. Osello
Polytechnic of Turin, Turin, Italy
e-mail: daniela.deluca@polito.it

C. Bolognesi and D. Villa (eds.), *From Building Information Modelling to Mixed Reality*, Springer Tracts in Civil Engineering,
https://doi.org/10.1007/978-3-030-49278-6_8

1 Introduction

The technology revolution, that has involved the last years, have generate the new system for the managements and maintenance in the innovative industrial process. The new vision of modern industry involves technological progress in various fields. For example, optimized systems for production management and planning make use of automated processes to obtain data from production lines and machines (Roblek et al. 2016).

The industry sector has invested significant resources to integrate intelligent systems and interconnections into its information system for management (MES). One system that guarantee the control of the process thank an interconnection between differ lower system is MES (Mantravadi and Moller 2019). This method is used for the planning of activity in the production, maintenance the orders, the response to random events and the adjustments of plans and the following-up of activities (Blanc et al. 2008).

During the time, the information system that different industries adopted has suffered computer implementations for reduce and simplifying the data loss (Mantravadi and Moller 2019).

Therefore, the installation of intelligent devices and local sensors helps maintenance and allows the factory to be digitized by turning components into intelligent resources. The element that characterizes the new production system is the integration between automation systems such as the Cyber Physical System (CPS) and the interoperability of the knowledge and skills systems within Industry 4.0 (Buchi et al. 2019).

The Cyber Physical Systems are defined as an enabling technology. By combining acquisition systems with enabling technologies, interoperability between systems is guaranteed to implement current industrial methodologies (Lee et al. 2015).

For evaluation of the new system it is necessary to define alternative methods to the classic way to represent the factory, using the virtual modelling, advanced artificial intelligence (cognitive) for process control, which includes autonomous adjustment to the operation systems (Rodic 2017). Thanks to these innovations, was introduce the concept of Digital Twin, extending the use of modelling to all phases of the product life cycle. The Digital Twin is defined as the virtual model similar to the real object, used for simulated and generated specific process and method. Once tested in the virtual setting, the information was transfered within the factory by comparing simulated data with real data (Negri et al. 2017). Given the great potential that virtual models of simulations outline within a process of optimization and innovation was possible to use the Building Information Modelling as Digital Twin becoming a large repository of alphanumeric and three-dimensional information.

In this contribution, the Digital Twin is used as a 3D virtual model, for simulated and processing the information came from external database. For this reason, the Digital Twin could become an emulator for evaluated different efficiency scenarios in term of time and cost in the industrial areas.

In this context an example is the Regional project DISLO-MAN (Dynamic Integrated ShopfLoor Operation MANagement for Industry 4.0), which aims the creation of an Information and Communication Technology (ICT) platform for the integrated, dynamic and autonomous management of production operations to optimize the use of resources for the creation of new connections between people, materials and production systems. In order to achieve the goal of the project, a virtual warehouse has been reproduced through parametric digital modelling, to which all the management data have been associated; for example, the average stock, the rotation index. This method has defined the correct workflow to connect different platforms and sharing data. After connecting the Building Information Model with the management system, data could be extrapolated to analyze and update them in real time through Mixed Reality applications. In this way employees could manage the operation without loss of data of products in stock in their department. Through headset, approaching the product with its own identification code, a series of useful operations appear to update the database, each time it changes stock and position within the building. By adding a virtual mini map, it was possible to identify and track the product within factory, studying the smartest time and route to apply to reduce inefficiencies related to the production chain.

The challenges brought about by the industrial revolution involve digitalization through new control systems that use Virtual, Augmented and Mixed Reality systems to facilitate the individual worker, reducing errors and management costs. In this direction, dataset and database become the focus of innovation, studying methodologies and systems that could preserve it, but at the same time make it accessible to all without compromising its integrity.

2 Case Study

The DISLO-MAN platform was born from the need to make local companies competitive with highly specialized personnel supported by Internet-of-Things (IoT) technologies in their various nuances. In particular, the project involves the use of wearable sensors, systems for the indoor location of operators and automated material handlers such as AVG.

To facilitate the acquisition of data from sensors located on machines, conveyors, goods in storage and operators, it was possible to create a shared platform in which the data flow in a raw way and then processes them with data cleaning systems (Apiletti et al. 2018).

Once reworked, machine learning software systems could analyse complex production scenarios and provide viable alternatives, optimizing human and economic resources. In addition, the platform provides for the integration of production systems with machine productivity parameter data, thanks to the sensors installed, to optimize the flow of materials and minimize downtime. With IoT systems, the user of the future users will have the opportunity to be guide by motion recognition systems to make their job even more ergonomic.

Through statistical and comparison methods, real-time processed data ware trained to create predictive models, in which digitization highlights best practices to be visualized through Augmented Reality and Virtual. In this way Smart Glasses, Cave, Immersive Training Systems, become excellent tools to evaluate the performance of human activities.

The benefit that the project adds to the local industrial system characterizes three fundamental aspects. The modularity applied to shop floor contexts of different nature, to create a dense network between sensors and data standardization rules with simplified access interfaces. Interoperability then plays a central role, thanks to integration with existing information systems. In this way, using the resources available on the market, interoperability compares the results of different sectors in a platform that unifies the data collected in a common language. Finally, the platform tries to use the same types of software to manage different hardware and visualization systems in different shoopfloors. In addition to these innovative aspects, through middleware, data analysis algorithms and machine learning, independent systems with dynamic integration created to manage and control the industrial process throughout the entire production cycle.

The opportunity introduced by the Dislo-Man project was strengthened by the local production system by involving many industrial partners, companies supporting the IoT and research institutions. In terms of numbers, seven large companies located in Piedmont and Valle D'Aosta, twenty-eight partners for the IoT, two research institutions (Polytechnic of Turin and University of Turin) and as many as one hundred and twenty people involved throughout the project. Started in October 2016 and ended in October 2019. At the end of the project, the impact of economic/management skills was assessing, increasing new business models and turnover for the SMEs involved. Within the innovative systems it is feasible to introduce the methodology of Building Information Modelling (BIM) to return in three-dimensional form the building and the sub-components, in order to determine also a database in which all the alphanumeric information associated with each element converge; e.g. relationships between different objects, materials, costs, construction phases, etc. (Osello 2012).

The use methodology in order to achieve the project's objective, optimized the entire production chain thanks to the creation of the BIM model that connects and facilitates access to the data processed within the Dislo-Man platform. Being able to create a unique parametric digital model, it could be shared among different users and keep its alphanumeric properties unchanged during the different simulation phases. The model hires the role of Digital Twin, useful to diversify different scenarios and implement data from internal information systems.

3 Methodology

3.1 Workflow

Through the identification of input needed in Fig. 1 to define the correct workflow, it is possible to connect different platform with many domains and data share. In the first phase of creation's Digital Twin model as Building Information Modelling is important to identify the parameters and metrics required to optimize the management of technical areas. For this reason, the model is based on the internal database came from by the industrial system information.

The data of virtual model could be transferred into free platform on web, using query system. In this way, more users and more instrument have to the possibility to access at the information obtained sharing the model.

What characterizes this process is the uniqueness of data, editable using Mixed Reality applications. Moreover, the actors of this method could compare many information with the real situation, so the process is validated. During the final phases the

Fig. 1 Workflow and methodology use for innovating the industrial case study

model with his information could be updated in real Time in the web platform and imported them into geometric model. Is important to connect the new model with the MES for generated a unique virtual system.

3.2 Digital Model as Building Information Modelling

The BIM model by its very nature, in addition to the geometric aspects, preserves, through associations of parameters, data from various sources such as technical data sheets, tabular values relating to the production and allocation of products, safety data sheets allowing an integrated approach with respect to the information that each element may contain.

In this way, by combining several paper archives in digital format, there will not be duplication of information but an immediate geometric and alphanumeric correspondence of the selected object. To obtain a good graphical database, it was useful to stable the appropriate level of detail (LOD) to ensure an adequate level of information and representation without losing definition and evaluate the appropriate visualization through virtual interfaces diversified according to the final goal. The acronym LOD varies according to the regulatory context in which the BIM model was developed. Among the definitions it is useful to mention the American Institute of Architects description given in the 2013 year. This level of development considers the quantities graphic of the information insert into BIM model, so in every object the information is select based on phases choose (Abualdenien and Borrmann 2019).

This definition clearly explains the need to establish minimum dimensions, spatial, quantitative and qualitative in an element of the model based on the uses (De Luca and Osello 2019).

In addition, a ranking was established with five levels from 100 to 500, which are associated with geometric and alphanumeric requirements. With an LOD 100, the model was conceived by means of generic masses without any level of detail, in the LOD 200 it is possible to define an approximate geometry of the elements, instead from the LOD 300–500 the level of modelling begins to be elevated until the obtaining of a Digital Twin identical to reality. Following the classification, the BIM model created follows three types of LOD. For the building envelope, the modelling complies with an LOD 400, while for the machinery, an LOD 300 was chosen, integrating the missing components through the insertion of related alphanumeric parameters.

Finally, for the warehouse and its products, the modelling follows a LOD 100 defining the volumes and spatial dimensions, leaving room for alphanumeric indexes. The peculiarity of this methodology is the great opportunity to create hierarchies of data easily manageable and interoperable with non-BIM platforms, through overlapping schedules. Through specific plugin, the BIM model was populated with useful information present in the MES system, selecting what to display in the geometric model. Once again, the interoperability, even if with small gaps, could be satisfied and the integration between different systems is guaranteed by web dashboards able

Fig. 2 BIM model of case study

to visualize the data. The more closely the model was aligned to the objectives, the more the reliability of the process meets the physical and functional requirements that the fourth technological revolution requires.

For these reasons, an example of a process of storing goods between different demonstrators was identified for to be replicate within a new container, through the main functional characteristics. This choice allows the system to be flexible and to generate data on a geometric level for further scenarios. The structure used to test the BIM methodology in the industrial field is the former complex of turning shop located near the Polytechnic of Turin. In Fig. 2 the virtual model created contains only the outer casing and the horizontals and vertical connections. The entire supporting structure of the roof has been preserved in order to evaluate, in subsequent studies, the logistical dimensions of the entire production line.

The choice to relocate the spaces comes from careful studies on the different production layouts, preferring a large space able to resemble in shape and size to the real model of the warehouse, the focus of this contribution. Based on the information received, the choice of placing the storage department in the central building was in accordance with the actual size and number of racks. By calculating the maneuvering areas of the forklifts, the maximum size of the racking and the number of openings required for the goods, the virtual warehouse could be considered to all intents and purposes to be similar at the real.

After allocating the different departments, the study focused on digital modelling of the warehouse. The maximum number of racks and pallets that represent the useful dimensions of the products, derives from the standard dimensions of the norm UNI (2004) UNI-EN 13698-1. There are various criteria for the classification of the warehouse loading units of the categories of units stored, the level of automation (manual, semi-automatic, automatic) and finally the status of the material.

By identifying the load units and following the instructions in the standard, the distance between the racks considers the means of handling such as forklifts or AVG that could help the operator in picking up and arranging the products. In this case, the minimum clearance guarantees a minimum maneuvering space of 1.20 m for three-way trolleys. Combining the different requirements, a storage model has been obtained: with variable containers allocated unevenly according to size, automatic automation systems for picking and direct control over the most fragile products. Each volume has been associated with an identification code, a specific coloring to display the position occupied and all the parameters useful for the correct management of the products, deriving from files linked to the MES. The products of the department, following the priorities of variable handling based on the index of rotation and type, need a clear coding that identifies all the parameters useful in the different phases of allocation (Rimella 2018).

The mapping of the products was carried out by identifying three classes of access to the articles, the layout of the shelving and the crossing policy provided, and the articles are positioned with the aim of reducing the main picking times: distance of the route, search for material and withdrawal of the codes indicated in the order (Graziadei 2004).

In fact, through the re-elaboration of the data, a matrix has been created. It contains the values of the average stock, the annual consumption of the product obtainable from the accounting sheets and the cost on the market of that specific element. (Rimella 2017) In this way, it was possible to obtain the rotation index that allows to correctly allocate the products. To facilitate the modelling, in Fig. 3 the virtual model is trained for the visualization, using specify script to calculate the tabular values.

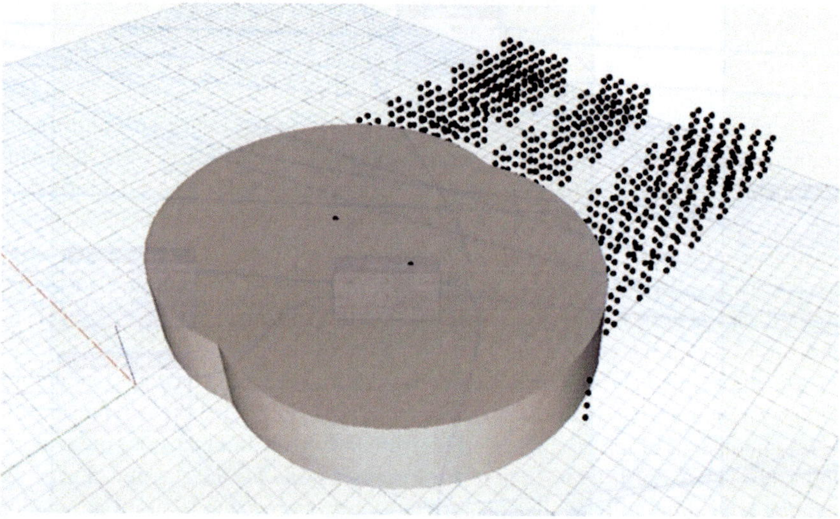

Fig. 3 Dynamo script to identify storage areas

Each object of the model has been associated with the alphanumeric identification code, the rotation index, the storage area and the condition (occupied/free). The Condition parameter, a material has been associated which, according to the zone, retrieves its status during the day. After having established the rotation index below which a product must be allocated in zone C and the index above which a product falls in zone A, it was necessary to size the different areas of stock so as to be able to draw every single movement of the product. The visual programming software Dynamo was used to do this. Through the programming, the comparison between the indexes inserted in the BIM model and the rotation index of the categories of articles reported on the external file, the model highlights the number of elements that should be contained in the single zones. After this is necessary to calculate the minimum radius that should have a circumference with the center of curvature in the middle of the exits in the warehouse. The need to calculate this value arises to define the width of zones A, B and C. These relationships extract the positions of the elements placed on the shelves of the digital warehouse; these allocations are read and represented in Dynamo as points of coordinates x, y and z. It was therefore useful to intersect the volumes of the cylinders previously created with the positions of the elements in order to identify exactly the objects in the zones. The comparison of lists was done to overlap heterogeneous data and special character using Python scripts.

Once the areas with elements of the model that are part of it have been sized, all that remains is to modify the parameters of the selected objects by entering the information provided in the external file. By varying the material with a new one previously created, it is possible to view the maximum and minimum capacity of the warehouse, whenever the external files could be modified. Thanks to the use of linked database, with special scripts the model receive the changes and they will adapt at the new indications. During this operation, sometimes an overcrowding or an undersized department emerges. In this way is highlighting the lack of some areas with insufficient space to accommodate all items. In the Fig. 4, the digital model, by means of an "alert", warns the operator of the problem encountered. Once again, through visual programming, it is possible to return a message indicating which area has been undersized and how many products remain outside the shelving. Following this methodology, the simulation model, returns different values and characteristics both at the level of the individual product and of a general management of the department.

3.3 BIM Parameters and Management Indices

The logistic system was defined as the set of organizational, management and strategic activities that govern the flow of materials from the purchase of raw materials from suppliers to the delivery of finished products to customers and the after-sales service.

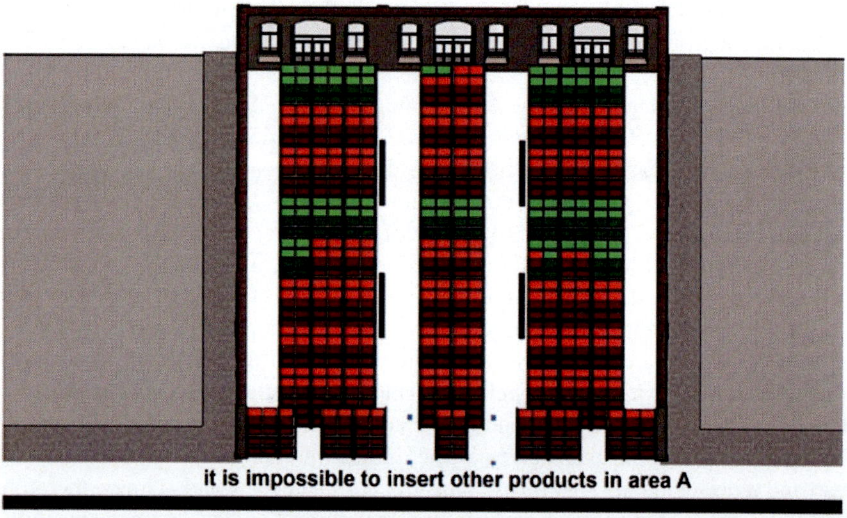

Fig. 4 Overcrowding of products in the selected area

Indices and parameters were defined that were able to improve the efficiency of the raw materials storage system, the correct allocation and elimination of the final product (Vignati 2002).

These indexes described in the previous paragraph have been inserted into the BIM model through scripts for each individual product. In Table 1 are shown the example of products with the specific index using for evaluating the efficiency of stockage. After this first operation the second step is to create for each object of digital model the camps featured in the external file. Using script, every information is reading and associate at the object that have those camps. This passage is automatically because the script recognizes the ID number for each object and read the camps associated.

To evaluate the correct use of data, the rotation index was used, highlighting which articles remain in the company and for how long. This is also classifiable according to the type of factory and products made; many companies that produce fresh products will have a high index while other types of industries are characterize by low values. In order to avoid goods occupying unnecessary space and risk being ruined by time, the turnover rate is the optimal solution to reduce losses. This index could be calculated by the ratio between the outputs of a period and the average stock in that period (Urgeletti Tinarelli 1981).

The reference period can have a variable size, referring to the single product, to groups of articles or to the entire warehouse. In the case study, the rotation index considers the single product present in the area during the whole year. Another parameter evaluated was the average stock the quality/value of the articles in the warehouse. These indices are added thanks to the management software, by entering the purchase data and the products deriving from the production line. Analyzing only these two factors, the virtual model developed could be considered a dynamic

Table 1 Management indices

Items	Class	Average stock	Annual consumption	Price (€)	Stock value	Consumption value	Rotation index
Enapren 20 mg	Class A	141	1831	11.31	1594.71	20,708.61	12.99
Aulin bs	Class A	219	1713	4.91	1075.29	8410.83	7.82
Ciproxin 500	Class A	68	298	14.4	979.20	4291.20	4.38
Dostinex 0.58 cpr	Class A	2	15	57.11	114.22	856.65	7.50
Totalip 20 mg	Class A	3	105	54.9	164.70	5764.50	35.00
Tavor 1.0	Class C	106	990	6.3	667.80	6237.00	9.34
Muscoril f.le	Class C	7	287	10.35	72.45	2970.45	41.00
Gentalyn Beta cr	Class C	7	317	13.97	97.79	4428.49	45.29
Xanax 0.5	Class C	14	265	6.83	95.62	1809.95	18.93
Glucobay	Class C	2	74	13.94	27.88	1031.56	37.00
Neolactoflorene fl	OTC	292	342	8.7	2540.40	2975.40	1.17
Aspirina C 20 cp	OTC	204	881	6.9	1407.60	6078.90	4.32
Tachipirina 250	OTC	74	414	4.1	303.40	1697.40	5.59
Moment 200 24 cp	OTC	77	395	7.8	600.60	3081.00	5.13
Maalox Plus sosp	OTC	60	76	5	300.00	380.00	1.27
Garze 10 × 10 × 100	Other Products	80	183	2.3	184.00	420.90	2.29
Euphralia coll	Other Products	27	151	6.8	183.60	1026.80	5.59
Mediker sh antiparass	Other Products	45	129	8.9	400.50	1148.10	2.87
Saugella derm.liq 500 ml	Other Products	29	53	9.3	269.70	492.90	1.83
Chicco Gommotto	Other Products	6	51	3	18.00	153.00	8.50

compact warehouse. The advantage was the reduction of infrastructure management costs, standard handling equipment and a high level of space saturation.

Since some items could remain in their storage area for a long time, the coverage index obtained from the rotation index, allows to assess the time of inefficiency of the system and understand the maximum demand for the article in the global market. Once the indexes were established and the areas of reference identified, each article was coded according to precise rules.

Each code contains classes, subclasses, subgroups, types, subtypes, articles, starting from the general category to the single detail. In our case, each component has a part in letters indicating the area of stock, the subclass in numbers and then the progressive that follows the position of the article. In this way, in addition to simple sequential coding, allows a range of values to be assigned at the sub-areas, facilitating the identification. If the coding is clear and unambiguous then there will be a good precision between the code and the article, so is possible apply the simplification in terms of length, adaptability to variations and additions. The last numbers indicate the maximum capacity of the subclass with a maximum of 99 combinations, or 99 items. By limiting the number of products, overcrowding or anomalies in the shelving system were reduce; coding becomes an additional control tool able to support the operator during daily operations. Through interoperable tools and platforms, the new information communication processes meet the aims of the project without compromising their integrity using the BIM methodology.

3.4 Data Sharing and Web Cloud

For to improve and automate the operations of inserting and removing articles from the digital model; the data entered through the procedure described above, must relate to an external relational database that could communicate updates via query. This operation is possible with special plugin on the modelling software exporting to DB Link. This interchange format allows exporting the various elements of the BIM model and the relative information in different system for the management of data, such as the MS Access platform.

In order to be able, the manager of information flow in the warehouse, it is necessary to set up some queries to improve the visualization and the manipulation of the data.

Using a primary key of the element ID code are extract from the category Generic Model all objects that have the attribute IDType equal to Pallet. Later the table created was divided according to the attributes of Situation in two additional tables that identify the busy workstations and those empty. By dividing the database into relational tables in this way, it is possible to proceed to the creation of forms that allow facilitating the manipulation of data and communicating with the values of the management system. Using the Microsoft Access platform, the interface highlight with the Fig. 5 could be customized with different parameters. The "Product insertion" form was able to automatically fill in the fields of Code, IR, Condition and Zone; in fact, these parameters depend on the type of article to be inserted, it was therefore sufficient to query this data and click on the "insert" button to fill in all the fields listed above in the correct way. The mask use query for to change the element properties that have as the parameter "Situation empty". A second mask created allows you to remove products from the warehouse instead. This mask is able to select the code of the article and remove and recompile the data of the entity if the user has changed the real conditions. Once the operations on the relational database

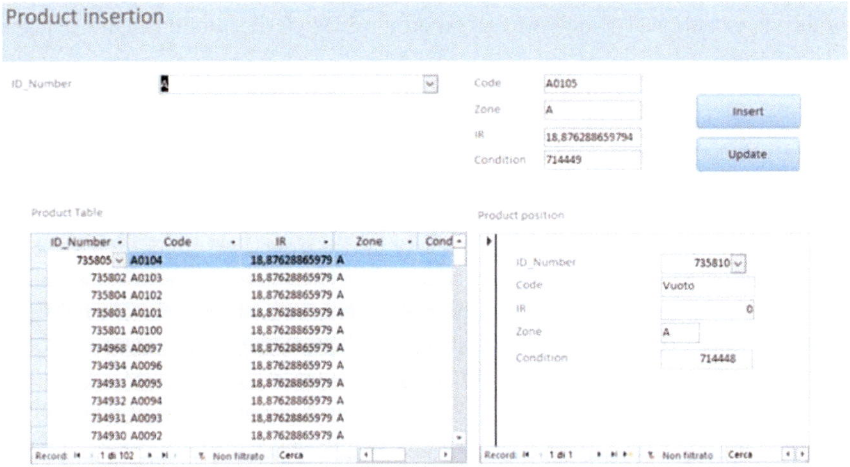

Fig. 5 Interface used for inserting products

are executed, it is possible to update the digital model using the DBlink plug-in and immediately see the changes made. Once again, this methodology allows to identify the discontinuities and verified the codes through simplified procedures.

The creations of this external dashboard for the data sharing, facilitated the connection on the web cloud. Indeed, the two masks transfer the same text file that the user established for the BIM model. Thanks many queries the virtual model contains all geometric and alphanumeric information's performed during the activity of the technical areas.

3.5 *Interoperability Between BIM Model and Application for Mixed Reality*

Today BIM authoring platforms doesn't guarantee a direct interaction with Augmented or Virtual Reality solutions, for these reasons it is useful test the interoperability process and apply the standardization. Analyzing the different software present on the market that are open source and compatible with the BIM methodology, the attention focused on Unity 3d. This platform was born mainly for create the virtual game and only in recent years it used thanks the versatility in importing parametric models. The problems for this program, concern the alphanumeric properties associated with three-dimensional objects in fact, it is difficult to associate the information without reassigning them through a programming language. The first operation to be carried out on the BIM software is differentiate the model through two database, one strictly geometric and one informative with all the data necessary for the visualization. The only way to export the BIM model into this platform is the FBX format.

This specific extension has the possibility to keep within the three-dimensional model, the elements' hierarchy, materials and complex geometries with their coding. By analysing the nomenclature that each object keeps, it is possible to verify the parametric family name, type and ID. In the first step, the information database must be structured considering the ID in order to export it correctly in the visualization software. The format that read correctly every information is the json. It is based on a Java programming language, that recognizing data in string, in Boolean, in variables and ordered sequence of values to which a function is associated to display additional properties. By supporting may coding such as text and number sequences, any name give at the parametric element is perfectly integrating into Unity. The information database exported is manage by Dynamo software to convert existing values into reading data with the JsonData format. Within this extension, the parameters reworked according to the selected data input, in this case study the ID number, the Condition parameter. Once the databases have been correctly importing through an appropriate script, each object is associated with a parameter corresponding to the ID and entered the necessary operations; for example, identify its position, update its condition. To speed up the system is necessary to create a list of items that to simplify the search and selection. The Unity software includes a Mixed/Reality application called Vuforia, able to associate targets or detected objects into the virtual model. Through virtual buttons, the user can interact with the Digital Model and become an integral part of the process.

4 Results

The Mixed Reality applications support daily worker's activities, meeting the challenges of the fourth industrial revolution (Quint et al. 2015).

Through headset and simple interactions, it is possible to increase and innovate the communication between different employees within the departments. It is really the optimization of communication and the experimentation of new management methodologies that transforms the tools into intelligent objects able to process large amounts of data and make them more accessible with a simple "click". This system highlights the advantages of Mixed Reality during visual inspection of intelligent objects (Jakl et al. 2018).

Data management system is a central element in a warehouse commissioning activity, as it needs to create a database that is updatable and easy to manipulate during the daily operations (Latif and Shin 2019).

Comparing different studies, it emerged the need to provide a specific function to modify directly from the application in MXR the information on the position of the products during the phase of procurement of resources.

The result of this job brought the identification of the new innovative methods for manage directly the data using Mixed Reality's headset. Indeed, the tools of

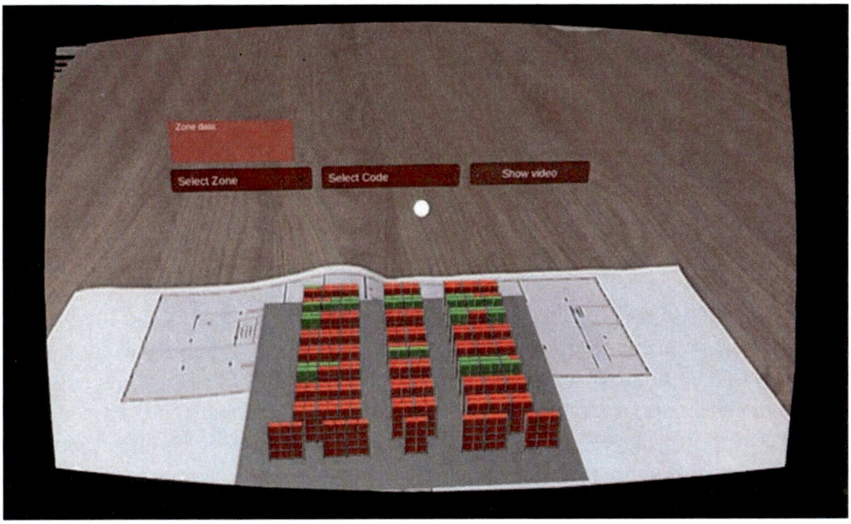

Fig. 6 Warehouse visualization in mixed reality

Augmented Reality and Mixed Reality have been used in this contribution for the goods receipt with respect to the delivery note and to decide where to place and how to place the articles in the specific areas. In Fig. 6 it's possible to view the position of elements and their conditions, for example if the pallet contain the article or not. Selecting the zone in the application, the script shown only element into this area, so the user is facilitated to search the object.

These new technologies can help the operator to quickly identify the product during the "picking" phase. During the mentioned phase, in addition to view the product, it is also possible to communicate errors during the tracking/moving of goods, scan the barcode and assign it to the picking cart or provide information to avoid situations of overcrowding (Stoltz et al. 2017).

Thanks to the great potential that emerged during the picking phase, the application in Mixed Reality developed tries to find innovative solutions to solve some of the needs highlighted above. The identification of individual products, follows the real localization, taking into the coordinates of single articles present into warehouse.

The sectorial nature of the areas of storage is based on the correct allocation of the product and the sharing of the 3d model and its database on cloud platforms define innovative method. The user could decide based on his role and the responsibilities assigned what to see, using simple buttons. In a first phase, the initial menu provides the possibility to view in video format, the entire construction process of the BIM model and the association of alphanumeric data from the management system. In addition, it is useful to analyse the effectiveness of parametric 3D visualization, during the monitoring phases of the components present within the area. In this way, the specialized employee could be trained without any human support, simply by following a video tutorial where in detail was explained the process of implementation of management of the BIM model. Through another two buttons, once on the spot it was possible to superimpose the entire digital model to the physical one and to interrogate through tags the allocation of the goods divided by sector and identification code. Selecting the "Select Zone" button allows you to choose the area of interest and view only the pallets and racking compressed within that area. So, the user analyses the number of empty or occupied positions on the reference day selected. This data could be linked to an online database that can be modified by all the employees in the department. The advantage of being able to compartmentalization the warehouse ensures greater reliability of data, better controlling the priority of the products present or arriving, without having to carry out a second control.

The second button inserted give at the user the possibility of displaying the single product in the area select, after it will light up on the virtual model and consequently, with the right over-lap with the real position, it will be easy to identify it. For this part is important to add all the information related the product and with any other commands update its position. By combining two visualizations, it was possible to make a further check on the paper information coming from the commercial area, verifying if the products were correctly positioned according to the order of exit. Specifically, in Fig. 7 the application allows to search for a specific product in the warehouse through a match between lists. Each table contains the list of all product codes in the department and could be compared with each individual attribute that is provided as input by the user. If the data are the same, then the match between the data is satisfied. The final output of the process makes able to extract an index used for to select the geometry corresponding and to assign the position. The ability to share the parametric model on the web and manage its properties through basic

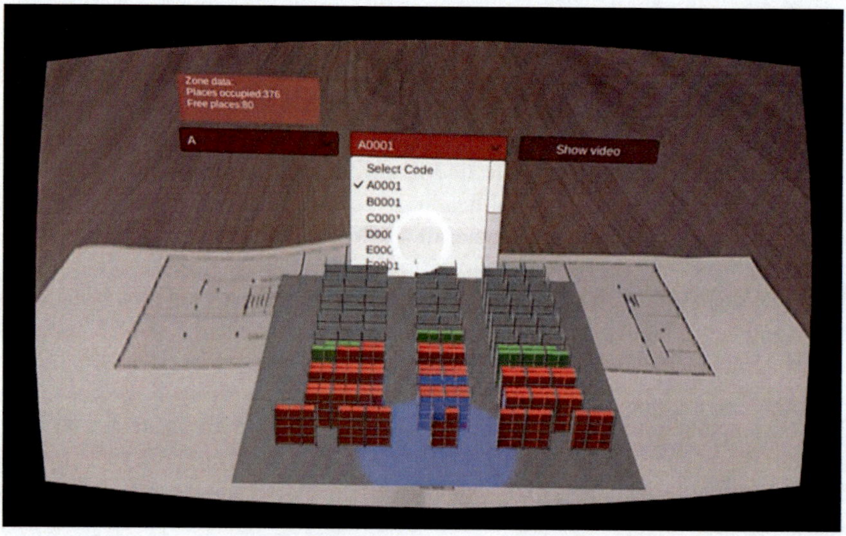

Fig. 7 Visualization of product into warehouse

interactions provides good support to different users who use the application in Real Time. Whenever the user deletes a product or changes its allocation priority to a shared database, the application updates this information and view new content. In this way, the virtual model facilitates any overlapping of elements and allows more users to manage in an innovative way the products even in times of great criticality or overcrowding.

Using multiple scripts created specifically to meet the needs at different stages, it is possible to recall the data of the devices in the field, cataloging and historicizing the problems in a single database. Since the first simulations, several inexpert users of these technologies have responded positively to use this new optimized warehouse management. The first results are related to the reduction of paper by leaving delivery notes and transferring the other documents to the cloud for editing in Real Time and visible by all users. The second advantage is the optimization of the storage time of the articles, as the application described above facilitates communication between paper documents and real object through the reading the identification strings. Furthermore, give the way to access at the information freely, it is possible to compare data from outside the industry or to manage the customer's request in an optimized way and to transfer correct information to other employees in the departments. In order to carry out this type of process, other partners have developed algorithms to convert the scripts used for data transmission in the departments.

Everything becomes traced in this way; creating a unique and accessible archive at any time of the day. Another important result achieved thanks to the DISLO-MAN platform is the degree of technological innovation that the companies have put in place. Every script and algorithm have been tested and stressed to the maximum level for to optimize the process also thanks to Mixed Reality applications.

5 Conclusion

By searching for advanced viewers and software it will certainly be easier to manage interactions and enter a large amount of information. Unfortunately, the limit of these technologies does not allow for a fluid and fully interoperable flow, but it was necessary to apply simplifications to the digital model or rework data, remaining as close as possible to reality so as not to change the simulated results.

The implementation of new technologies such as Mixed Reality headset directly connected to the BIM model will increase the productivity and efficiency of the factory system. In addition, new forms of investment linked to production optimization through direct control of operations will increase and reduced the human errors due to poor training and prevention. Innovating means optimizing, digitization involves a greater investment in methodologies and tools suitable for simplifying the management of a complex system both in terms of human resources and for the protection of critical data.

Increasing the BIM methodology, so far widely applied to the world of construction, in complex sectors such as Industry 4.0 means experimenting with tools that could connect the real world to behaviours and virtual resources increasingly similar to each other. The connection between these two apparently distant worlds will only happen thanks to a cultural and social change, which over the years will involve not only the industrial world but will lead to a new vision of society, going beyond current innovations. The new society 5.0 based on increasingly smart and interconnected communication methods (Fonseca 2018).

A viewer will no longer be enough to increase the perception and information that surround us but through avatars, we will be able to control in Real Time more sectors. Creating a shared database also allows users to implement applications with push notifications as soon as the files change so that inefficiencies can kept under control. In this moment, the applications in the industrial field that use the integration BIM-MES and VAR tools are not many.

Acknowledgements All the authors are pleased to thank the student Nicola Rimella for authorization to expose his work started before through an internship related to the BIM methodology applied to a new warehouse management. The work done previously has outlined further developments described in his thesis, through brilliant work on the application of new technologies of Mixed Reality to optimize the entire production chain. For this brilliant work, the authors have decided, together with his valuable collaboration, to produce this contribution. In addition the authors thank the POR 2014–2020 of Piedmont Region (Italy), FESR (European Fund of Regional Development), and MIUR (Ministry of Euducation, University and Research, Italy) under the program "Fabbrica Intelligente" (Smart Factory), Action 3, "DISLOMAN" project, Dynamic Integrated Shop fLoor Operation MANagement for industry 4.0.

References

Abualdenien J, Borrmann A, (2019) A meta-model approach for formal specification and consistent management of multi-LOD building models. In: Advanced engineering informatics, Elsevier, vol 40, pp. 135–153

Apiletti D, Barberis C, Cerquitelli T, Macii A, Poncino M, Ventura F (2018) iSTEP, an integrated Self-Tuning Engine for Predictive maintenance in Industry 4.0. In: IEEE International conference on parallel and distributed processing with applications, ubiquitous computing and communications, big data and cloud computing, social computing and networking, sustainable computing and communications 2018, IEEE Computer Society, pp 924–931

Blanc P, Demongodin I, Castagna P (2008) A holonic approach for manufacturing execution system design: an industrial application. Eng Appl Artif Intell 21(3):313–330

Büchi G, Cugno M, Castagnoli R (2019) Smart factory performance and Industry 4.0. In: Technological forecasting and social change, Elvesier, vol 150, Article 119790

De Luca D, Osello A (2019) Strumenti e metodi BIM per l'Industria 4.0. In: BIM VIEWS Esperienze e scenari, Cua, Napoli, pp 21–32

Fonseca L (2018) Industry 4.0 and the digital society: concepts, dimensions and envisioned benefits. In: Proceedings of the 12th international conference on business excellence (PICBE) 2018, DE GRUYTER OPEN, pp 386–397

Graziadei G (2004) Gestione della produzione industriale. Strumenti e applicazioni per il miglioramento delle performance, Hoepli

Jakl A, Husinsky M, Schöffer L, Wagner M (2018) Augmented reality for Industry 4.0: architecture and user experience. In: 11th forum media technology 2018, At St. Pölten, Austria

Latif UK, Shin SY (2019) OP-MR: the implementation of order picking based on mixed reality in a smart warehouse. In: S.Y. Visual computer, Springer, vol 11, pp 1–10
Lee J, Bagheri B, Kao HA (2015) A cyber-physical systems architecture for Industry 4.0-based manufacturing systems. In: Manufacturing letters, Elsevier, vol 3, pp 18–23
Mantravadi S, Moller C (2019) An overview of next-generation manufacturing execution systems: how important is MES for industry 4.0. In: 14th Global congress on manufacturing and management (GCMM-2018), procedia manufacturing, Elsevier, vol 30, pp 588–595
Negri E, Fumagalli L, Macchi M (2017) A review of the roles of digital Twin in CPS-based production systems. In: 27th International conference on flexible automation and intelligent manufacturing, FAIM2017, 27–30 June 2017, Modena, Italy, procedia manufacturing, Elsevier, vol 11, pp 939–948
Osello A (2012) Il futuro del disegno con il BIM per Ingegneri e Architetti. Gangemi Editore, Palermo
Quint F, Sebastian K, Gorecky D (2015) A mixed-reality learning environment. In: International conference on virtual and augmented reality in education 2015, procedia computer science, Elvesier, vol 75, pp. 43–48
Rimella N (2017) Progettazione e gestione di un Layout di magazzino. Report finale di Tirocinio, Politecnico di Torino, pp 1–7
Rimella N (2018) Visualizzazione del modello BIM per l'industria 4.0 attraverso la Mixed Reality, Tesi di Primo Livello in Ingegneria Edile, Politecnico di Torino, pp 1–32
Roblek V, Mesko M, Krapez A (2016) A complex view of industry 4.0. SAGE J 6:2–11
Rodic B (2017) Industry 4.0 and the new simulation modelling paradigm. Organizacija 50(3):193–207
Stoltz MH, Giannikas V; McFarlane D, Strachan J, Um J, Srinivasan R (2017) Augmented reality in warehouse operations: opportunities and barriers. IFAC-Papers OnLine 50(1):12979–12984
UNI (2004) UNI-EN 13698-1, Specifica di prodotto per pallet—Parte 1: Specifica di fabbricazione per pallet piatti di legno 800 mm × 1200 mm
Urgeletti Tinarelli G (1981) La gestione delle scorte: organizzazione, contabilità e automazione. Milano, Etas Libri
Vignati G (2002) Manuale di logistica. Milano, Hoepli Editore

AIM! Algorithmic Information Modeling: New Strategies for a Fully Integrated Approach in the Field of Cultural Heritage

Massimiliano Lo Turco, Michele Calvano, Elisabetta Caterina Giovannini, and Andrea Tomalini

Abstract Algorithmic Information Modelling (AIM) is an acronym that can be used to discuss and reason on the relationship between algorithms used in Visual Programming Language (VPL) platforms and the Information Modelling processes. In the Architecture, Engineering and Construction fields (AEC) the data management related to those processes are more usually managed through BIM tools able to handle heterogeneous data set characterized by different disciplines domains. The use of BIM processes applied to historic buildings (HBIM) moreover introduces new needs for the management of 3D information modelling. The research critically investigates some case studies concerning the relationship between the use of BIM processes applied to different architectural heritage fields, from the (virtual) reconstruction of existing buildings to the improvement of the efficiency of design activities for management applications. Most of the activities involved researchers of the Representation field, jointly with the Restoration, Geomatics, Building Physics, Architectural Technology researchers.

Keywords Heritage · BIM · VPL · Algorithmic Information Modelling

---

M. Lo Turco · M. Calvano · E. C. Giovannini · A. Tomalini (✉)
Department of Architecture and Design—DAD, Politecnico di Torino, Castello del Valentino, Viale Mattioli 39, 10125 Turin, Italy
e-mail: andrea.tomalini@polito.it

M. Lo Turco
e-mail: massimiliano.loturco@polito.it

M. Calvano
e-mail: michele.calvano@polito.it

E. C. Giovannini
e-mail: elisabettacaterina.giovannini@polito.it

C. Bolognesi and D. Villa (eds.), *From Building Information Modelling to Mixed Reality*, Springer Tracts in Civil Engineering,
https://doi.org/10.1007/978-3-030-49278-6_9

1 Introduction

The topic of data interoperability is certainly central within the current operational strategies related to the AEC sector. The research faces the challenge to dynamically relate some of the most important software platforms that are used in the architectural field today focusing on their application in the field of cultural heritage.

Within an extremely generalized context, the experiences conducted by the research group here described, refers to the need to develop the most efficient approaches of graphic and alphanumeric modelling of BIM systems and applied them to historical buildings using the recent nodal programming tools, better known as Visual Programming Language (VPL). Within these processes, the parametric term becomes central assuming different meanings.

In BIM software applications parameters are key properties of the building elements and are central to the system to create dimensional links among various instances; in programming code, new parameters can be introduced as further variables of new phenomena, and this is a fertile ground for experimenting and studying totally new phenomena occurring to the building during its entire life cycle (Fig. 1).

New disciplines are characterized by different languages and processes that need a new set of data and information. To enrich and manage this kind of data into BIM environment became crucial the use of VPL suitable not only to manage parameters and attributes but also to generate specific geometries that need to be readable by BIM software augmenting the complexity of the workflow towards an interoperable and fully integrated approach. Algorithmic Information Modeling and its acronym,

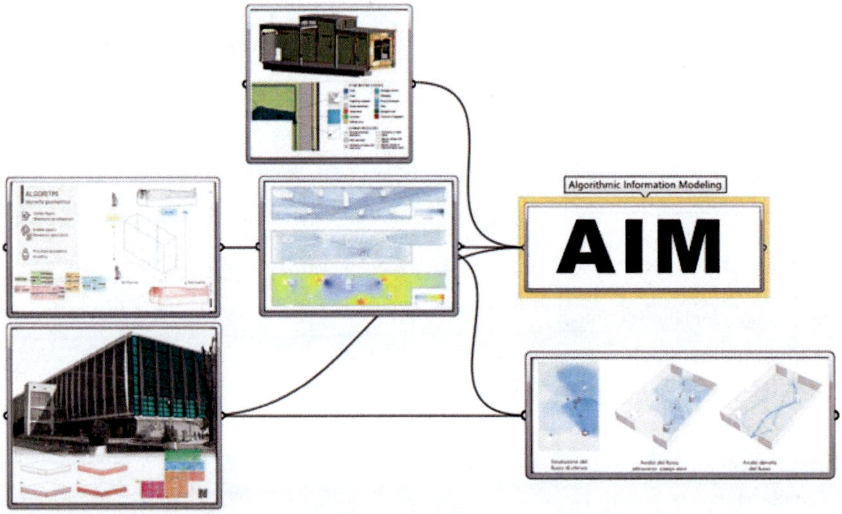

Fig. 1 When can we talk about AIM? With reference to the reported case studies, this methodology is introduced in cases where it is necessary to set up interoperable procedures between the BIM and VPL platforms

AIM can also be understood as the primary objective (the aim) to further develop interoperable processes that allow to unhinge some of the rigidities of the most widely used software platforms. Algorithmic procedures in Building Information Modeling are related with three main research topics: 3d modelling, information modelling and interoperability for the use and re-use of data.

The following case studies have the task of highlighting operational strategies and critical issues found in the resolution of different problems applied to extremely heterogeneous operational scenarios and application contexts.

2 BIM-Oriented Strategies for Cultural Heritage

In the architectural field, there is a wide diffusion of BIM applications, with particular reference to new building interventions. However, there are many criticalities if we consider the conservation field, for the enhancement and the management of the architectural heritage. Over the years it became necessary to critically reflect on the processes of instrumental information acquisition, standardization and data organisation, and their conversion into semantic models that must be interoperable and re-usable by different professionals in an interdisciplinary environment (Brusaporci 2017).

Starting from the definition by Murphy that describes Historic Building Information Modelling (HBIM) as a novel prototype library of parametric objects, based on historic architectural data, in addition to a mapping system for plotting the library objects onto laser scan survey data (Dore and Murphy 2012) the first investigate field in the past was to define best methodologies to obtain tridimensional models from a survey.

The metric acquisition phase of cultural heritage employing laser scanning and/or Structure from Motion photogrammetric techniques is concerned. This is the common starting point for the subsequent activities of reconstructive digital modelling oriented to the conservation of the historical-architectural heritage. Although data collection techniques are now very efficient and automated, the subsequent data processing still takes a long time to develop accurate models.

As far as the BIM methodology applied to Cultural Heritage is concerned, today there are still a limited number of research experiences aimed at understanding its potential in the specific field of historical buildings (Santagati and Lo Turco 2016). The theme is therefore still a frontier issue. The first difficulty concerns the lack of specific components/instruments for the modelling of historical architecture available within commercial platforms. In this respect, the reconstruction of complex and non-standardized forms still seems to be a particularly onerous activity (Bianchini et al. 2016).

In literature, some interesting works illustrate different approaches that adopt different applications to convert the point cloud into intelligent parametric objects, introducing the concept of the Level of Accuracy and Reliability (Bianchini and Nicastro 2018; Maiezza 2019). The most complex issue refers to the possibility of preserving the metric accuracy captured by laser scanners and photogrammetric point clouds even in the infographic modelling environment. Some research make a comparison between the point cloud and the 3D model to quantitatively assess whether the deviation between the numerical model and the mathematical one is within acceptable values (Fai and Rafeiro 2014). Some approaches prefer to perform 3D modelling in other platforms (also using procedural modelling based on VPL-Visual Program Language) to better generate and manage NURBS surfaces (which better approximate the trend and irregularity of complex surfaces), and then adopt operational protocols to convert NURBS into parametric surfaces that can be shared with the most common BIM platforms. The most productive field of application in this context is currently the field of classical architecture that offers an already standardized parameterization of architectural elements (Quattrini et al. 2018; Giovannini 2017; Paris and Wahbeh 2016). Other studies can be traced back to a "rigorous BIM", i.e. the adoption of the BIM methodology in the field of historical construction not only in terms of geometric precision, but also considering other variables typical of an information system (parametric objects, relations, attributes, the correct definition of graphic detail level, these are variables classified according to Italian laws in Level of Geometry—L.o.G.—, for what refers to graphic attributes and Level of Information—L.o. I.—for what refers to alphanumeric attributes). This work aims to reason on and explore the capabilities of historical building information modelling (H-BIM) for historical building restoration, to effectively combine with the geometric accuracy of the survey with the parametric flexibility and richness of information typical of the BIM processes, also investigating innovative interoperable processes with the algorithmic modelling platforms typical of VPL systems.

3 Between Heritage BIM and VPL

In recent years the research group has explored multiple approaches in an attempt to identify the best operational strategies to obtain geometrically reliable models that are prepared to accommodate and collect information equipment that can be used for different purposes, from documentation to monitoring, from restoration to functionalization. The proposed workflows investigate the use of BIM platform and VPL software to approach the restoration interventions management of historical buildings.

3.1 *Palazzo Sarmatoris: Adaptive Models for a Degradation Map*

Palazzo Sarmatoris, a seventeenth-century historical complex recognized in the engraving of the Theatrum Sabaudiae, located in Salmour, a settlement a few kilometres from Fossano (Piedmont). The ancient palace, currently owned by the "Casa di Riposo Villa Smeralda", although it is protected by the ministerial bond, is abandoned and in complete disrepair. Because of its deep stratifications, and the lack of diachronic care, it is in a widespread state of dilapidation, with instability and degradation that could cause the complete collapse of the entire structure.

The work consisted in the definition of a workflow that could be shared between different disciplines (Geomatics, Representation, Restoration) that in the first phase was characterized by the acquisition of geometric data of the artefact, operating an integration between different surveying techniques: point clouds acquired through terrestrial and aerial photogrammetry were combined with the point clouds acquired with LiDAR systems.

This was followed by the realization of the parametric model conducted in different ways, operating critical approaches of segmentation and interpretation of the numeric resource. Considering the state of conservation of the structure, the last phase consisted in the construction of specific parametric components able to highlight the visible deterioration of the structure, facilitating the historicization of the data and the possibility to mark, describe and group together all the areas with similar problems (Fig. 2).

Moreover, the quantity take-off of the decays and the required actions were summarized in the form of schedules. This methodology makes the current practice more efficient: usually, it consists of a simple graphical representation of the degradation phenomena without any connection with the information apparatus that describes the nature and the related procedures of intervention. In this sense, the management in the BIM environment is undoubtedly a more efficient system to storage both graphic and alphanumeric data, interrelated and implementable: this is a critical support to the conservation process, from the "event" of the restoration project/construction site to the "duration" of the maintenance/conservation processes.

However, although the procedure provides for a better correlation between data of different nature, the process of interpretation and definition of the various degradation phenomena is not entirely easy: the expert draws point by point the perimeter that characterizes the pathology, making the procedure extremely time-consuming.

3.2 *Bonavalle Castle: Algorithms for a More Automated Decay Representation*

Bonavalle Castle is a case study similar to the previous one in terms of morphological complexity and conditions. The castle, a medieval building, is nowadays in a state of

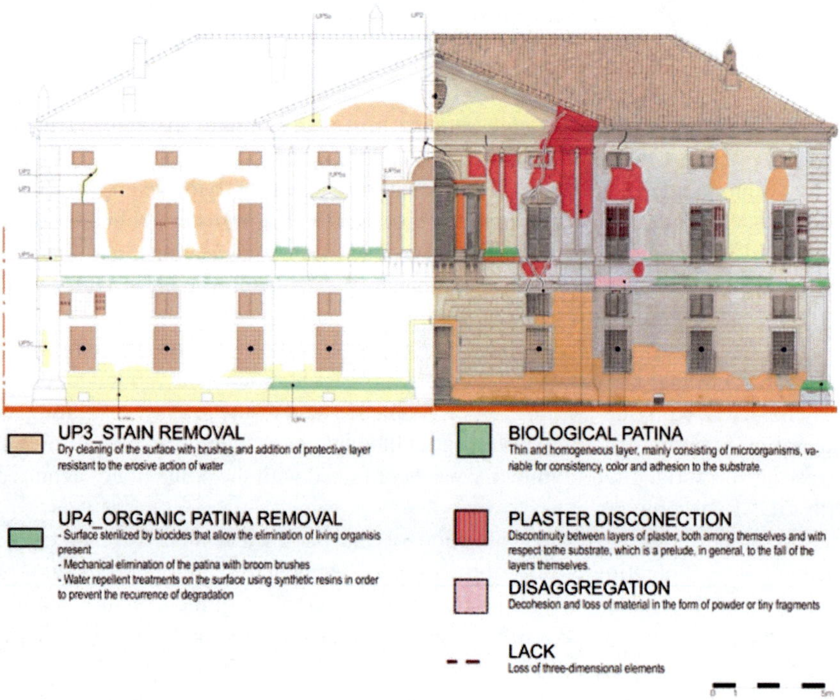

Fig. 2 Graphic and alphanumeric representation of the decays in H-BIM environment, through the use of parametric components. Palazzo Sarmatoris, Piedmont

disrepair. In the castle walls, it is possible to notice stratifications that can be traced back to different eras. The building is an example of what is called a "rural" castle, a settlement with an original defensive function, at the centre of a feud consisting of agricultural land, with a documented function of protecting people and products of the land. Over the years the castle losses its fortification functions, but it has maintained its characteristic function as an extra-urban "noble" residence, without definitively losing its original character, contrary to what happened in many other buildings that were transformed into villas from the seventeenth to the nineteenth century.

It is possible to identify two main phases related to the modelling of the castle and its decay. The first phase, related to the modelling phase of the castle was developed into a BIM platform, using existing tools, the second phase, for the creation of the decay map, involved also the use of VPL software to create algorithms that automated generate decay elements starting from orthophoto.

In order to digitally recreate the irregular sections that characterize the vertical components of the building, different conceptual masses have been modelled. The wall element that best approximated these sections was created using these masses as reference. To increase the level of metric accuracy of the digital model, once the virtual reconstruction of the principal elements was ultimate, a second phase was to proceed on inserting other architectural elements, both functional and decorative, parametrically modelled to better approximate the dense point cloud. The work continued by modelling the state of conservation of the building through the mapping of the decay. As far as the representation of degradation and disruption is concerned, different approaches were tested in order to obtain better accuracy in the shortest possible time, through the use of VPL algorithms.

The experimentation has investigated the potential of VPL through the use of Grasshopper and Dynamo software, with the main objective to achieve a greater level of automation through the use of some digital products available, such as orthophotos or two-dimensional drawing processed in dwg format.

The use of Grasshopper to generate algorithms, it allowed, at first step, to generate new surfaces containing the decays starting from orthophotos.

After a colourimetric modification of the orthophoto was possible to distinguish different type of decay and generate their perimeters. Polylines generated by the algorithm were converted into surfaces and classified and generated in a semi-automatic way through the use of the Grevit plug-in (connecting Grasshopper and Revit). The creation of real components inside the BIM platform allowed to develop an interoperable process between the visual programming language and the parametric environment based on building objects.

One of the challenges was also to face the problem of accuracy of the tridimensional model compared with the irregularity of the point cloud by was generated the orthophotos. Trying to fill the gap generated by the partially overlapping of the two models, the proposed solution chose to project the surface of decay. The operation of projection was developed through the use of Dynamo, a second VPL- based software, and allows to identify the decay starting from the point-clod orthophotos and then to project it on the building object surface. The same methodology can be applied using different starting data. For example from a degradation analysis processed in a CAD format, it is possible to transfer generated surfaces into a BIM platform, then the digital model of the decay can be enriched with different types of information. In conclusion, the proposed procedure is, therefore, a more structured way to implement automatic recognition processes of surfaces affected by degradation, working exclusively with the chromatic alterations present in the images in true form extracted from the point clouds. Other non-surface based pathologies such as gaps and disruptions will necessary follow different procedures that must be developed jointly with other professionals that provide a particular knowledge of the historic building and reading of involved phenomena (Fig. 3).

In conclusion, it is worth remembering that the use of BIM platforms makes possible to carry out multiple analyses, using a multi-criteria approach defined through qualitative and quantitative data. The database linked to the BIM model can be populated by much other information, including the corpus of historical iconographic sources available in specific archives, the retrieval and classification of recent and vintage images and data related to previous interventions that are necessary references for the planning and management of future interventions (Chiabrando et al. 2017).

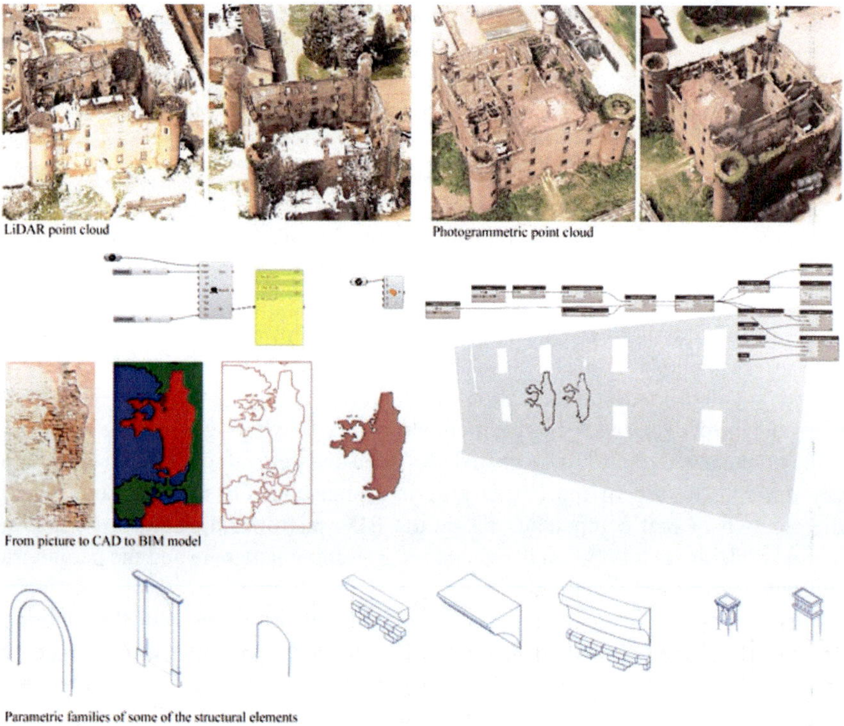

Fig. 3 Algorithmic procedure: from orthophoto to automated construction of closed profiles identifying the surface affected by the decays. Bonavalle Castle, Piedmont

4 Management in Cultural Heritage Field

Even if the BIM environment was first developed to manage processes about new buildings, in recent years, it has also been used for the management of historical buildings, generating new fields of research, involving also Cultural Heritage (CH) domain where the contribution of the algorithmic approach can help to create new geometries to make the current digital reconstructive modelling processes more efficient and interoperable. The adoption of VPL system is crucial for the planning of a new intervention on historical architecture, because of a wider capability to be enriched by information and data by multiple actors. In the field of museums, VLP can be applied to manage museum building and their collections, creating a common language for data interoperability between the exhibition rooms (container) and the artefacts (content). The following case studies, describe some developed workflows in these fields (Fig. 4).

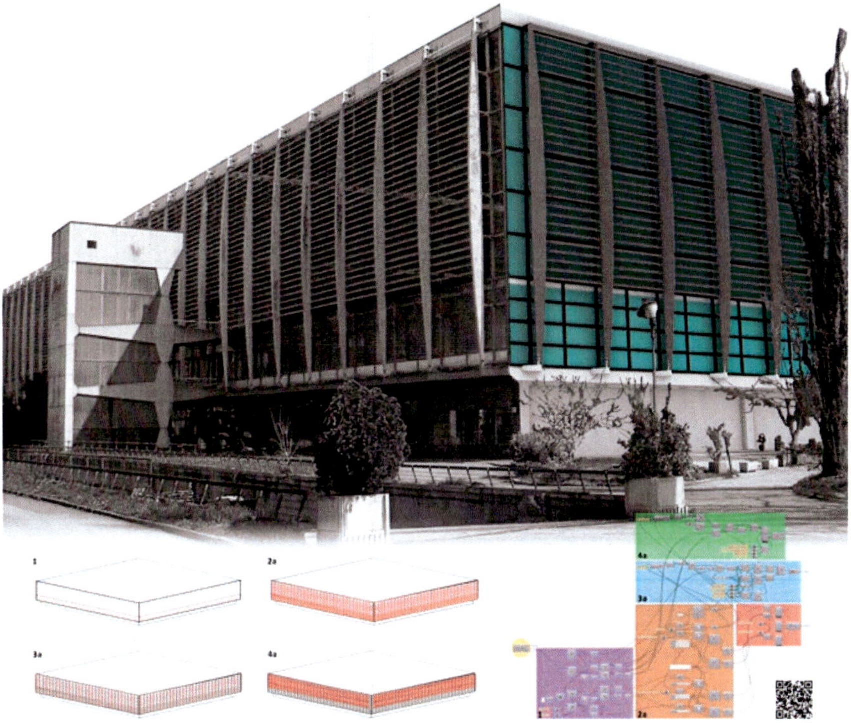

Fig. 4 Above, photo-insertion of the curtain wall modeled with VPL systems. Below, explanation of the sections that make up the parametric code. Palazzo del Lavoro of Pier L. Nervi

4.1 *Palazzo del Lavoro of Pier L. Nervi: Parametric Geometries for Automated Simulations*

The integration between these two parametric platforms has not only focused on the need to make the information heritage associated with the most recent integrated survey experiences richer and more controlled. The research group has developed new ways of sharing information useful to plan in greater detail the decision-making processes put in place in the subsequent phases of intervention on the historical building.

This is the case of the *Palazzo del Lavoro*, a work created to commemorate the Centenary of the Unification of Italy and designed by P.L. Nervi. Among the buildings made for the “Italia ‘61” celebrations is the one that, for its size and technological innovation, was the most published work by the main national and international magazines of the time (Pace et al. 2005). This masterpiece of modern architecture, through the 16 “umbrella” structural elements covering the 22,500 square meters of roof, the isostatic beamed ceiling of the perimeter balcony and the curtain wall system that describes the vertical curtain wall, is certainly one of the best representations of the dialogue between structure, technological renewal and architecture. Unfortunately, once the event was over, the City of Turin was unable to make the most of it and today it is in a state of decay.

Also in this research, the global scale approach provided for the realization of the HBIM model of the entire building, while some building components of particular importance for the subsequent reuse of the asset were modelled in VPL environment: the large curtain walls that characterize the envelope, and the technological subsystem consisting of uprights, panels and shading slats were investigated in greater detail to make explicit the geometric relationship between the parts and the related control over the daylight factor that permeates inside the building (Tomalini and Lo Turco 2019).

The lighting engineering evaluations were therefore conducted in a VPL environment. Depending on the results of the simulation, different design scenarios can be hypothesized that can be reported on the BIM environment through interoperable processes between the two platforms.

4.2 *SMART Museum Project: Toward an Integrated Collection Management System*

The SMART Museum project is the result of the collaboration between the research group and the Fondazione Museo delle Antichità Egizie di Torino, a museum characterized by important collections set up in equally interesting environments. Although it has about 10,000 square meters of exhibition space, many exhibits are still stored in depots waiting to be exposed.

The aim (here intended as purpose) is the connection between the three parts of a museum: the building (the container), the collections (the content) and the visitors. Usually, the relationships between these three themes can be shown through the exhibition; the layout proposed by the curator to allocate the collections in the exhibition space and design the paths suggested to the visitors (Ippoliti and Albisinni 2016). As mentioned above, specialized software allows to create architectures combining architectural objects in which dimensional parameters and attributes are included. Therefore, architecture can be represented by 3D digital building components plus data. The same approach can be applied to the collections: this means that a sequence of procedures and tools can be identified for the digitization and the data enrichment of the artefacts contained in the museums, which can be identified by the acronym Collection Information Modeling (CIM).

After having mapped the information contained in the BIM and CIM models, and having identified comparable common parameters in a digital environment, the curator can match the setting up of the acquired objects, starting from very simplified Building Physics controls, such as temperature and humidity. The informed representation allows the connection of information between the container and the

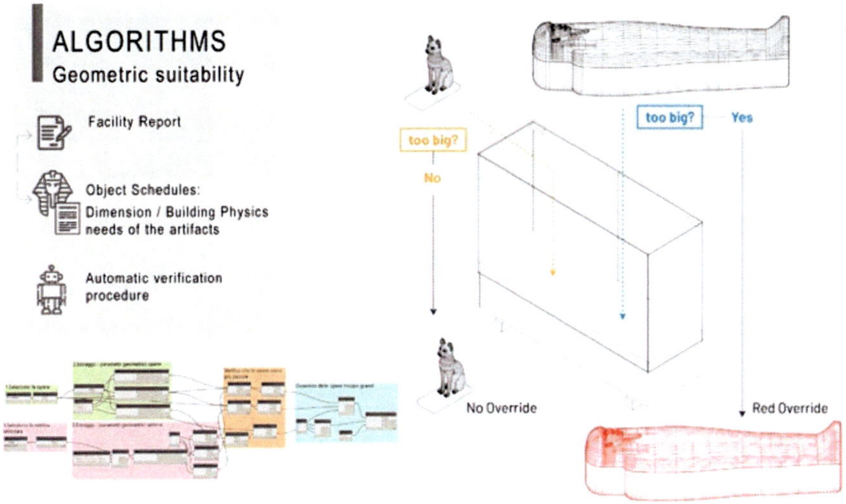

Fig. 5 The exploitation of the automated verification between the exhibition space, the dimensions of the artifact and the building physics features

content. The interaction between data can highlight a condition of the well-being of the artworks within an exhibition environment.

Therefore, once the conditions of well-being have been established, it is useful for a curator to think about the paths of the users within the exhibitions, highlighting the critical points where the flow of users can have some problems, such as bottlenecks and dead ends (Fig. 5).

So we can compare the artefacts as sort of attractor points used to evaluate the magnetic weight. For this reason, we must identify the parameters to understand the attractiveness of the object within the museum rooms. Some proposed fields are: the dimensions, the number of times that the object has been mentioned in web research, the number of people which saw the artefact in a previous exhibition, the historical and artistic significance just to mention the main ones. In order to better understand the research activities carried out, it is necessary to define some variables that have been useful to relate the physical environment and the collections.

In physics, a force field generated by a point entity describes the presence of a force applied to each point in a geometrical domain; the intensity of the charge is a parameter defined in relation to the characteristics of the associated metadata: in other words, points with high charge values generate a wider attractive force field. The association between charge and metadata of a point allows quantifying the "attractive weight of an object" within an environment, creating a hierarchy of contents. The decay represents the speed at which the effects of the charge decrease within the field. The visual acuity is the ability of the human eye to perceive as much detail as possible. The visual ray is similar to the sharpness of vision, therefore the ability to perceive distant objects. This parameter is set by the size of the length of the visual ray, and it allows you to define the amount of space perceived by the user (Fig. 6).

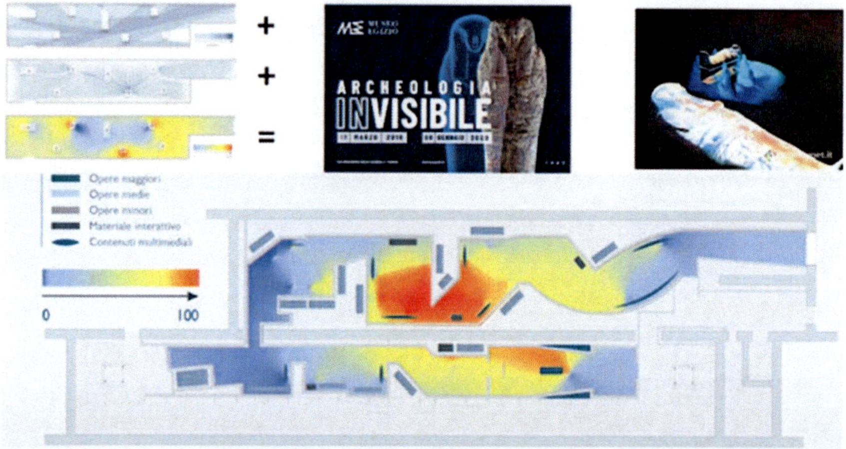

Fig. 6 Thematic maps that graphically describe the level of attractiveness and visibility of the collections

The prefigured behavior of the visitor derives from the adoption of the classification of the visiting styles elaborated in the ethnographic field by the scholars who grouped the different types of visitors into four categories, taking into account the amount of time dedicated to the visit, the movement in the museum exhibitions and the attention paid to the single works. This can be done by using known museum visiting styles to classify the visiting style of visitors as they start their visit (Zancanaro et al. 2007). They identify four different styles: the Ant, the Fish, the Butterfly and the Grasshopper (Veron and Levasseur 1983).

The on-going result is the creation of thematic maps that graphically describe the level of attractiveness and visibility of the collections. At the regard, the case study is the temporary exhibitions of the Museo Egizio named Invisible Archaeology is not so simple to be studied because it is characterized by a merging between artefacts and digital contents. The work is not intended to replace the experience of the professionals, but it aims to be useful support for the work of curators and registrars.

5 Conclusions

Thanks also to recent regulatory dictates, in recent years the professional practice has begun to adopt the BIM methodology mainly for large construction projects, in particular for new construction projects. Italian recent standards requests represents an acknowledgement of the widespread of BIM on a supranational level, especially due to its clear accordance with international standards and guidelines, regarding the crucial theme of the relationship between information content (graphic and alphanumeric) and construction processes (Chiabrando et al. 2018). The adoption of these methodologies on historical buildings is undoubtedly more complex (Volk et al. 2014). Furthermore the new UNI 11337:2017 standard, defining the different LOD for restoration, has been recently published and states that the restoration activities on existent construction considered of cultural interest imply a continuity of information management, and thus are based on data content and representation detail as resulting from the scientific conclusions made by the previous intervention (LOD F) implemented in successive management activities (LOD G). Nevertheless, the high level of detail, both for digital representation and information, is an essential trait of information modelling applied to cultural heritage. The particularities eventually insignificant for the single technical activities, in fact, may acquire a crucial relevance for other design choices.

At the moment, the processes and methodologies illustrated require significant and demanding data acquisition and processing time, which unfortunately the professional world, at the moment, does not dispose. This is the reason why for the research environment is crucial to shall actively cooperate with the industrial sector for a systematization of automatic processes and for the semantic recognition of data to reinforce possible interoperable connections. The research activities proposed in the contribution go precisely in this direction, towards greater process efficiency pursued through new and possible collaborative scenarios. Even if they are heuristic approaches, it is easy to realize that, experience after experience, the proposed workflows allow to obtain new and interesting results in order to organize more and more interoperable workflows in favour of process transparency.

The examples demonstrate how VPL can generate new parametric approaches for the construction of digital models and their management in some specific fields such as the integrated survey for subsequent intervention operations, the foreshadowing of design scenarios involving different disciplines and knowledge concerning the building physics and its components, or the relationship between informed models of museum buildings and the collections they contain make it possible to interconnect usually independent information systems.

AIM processes finally generate new set of algorithms to challenge different issues creating a new catalogue of available procedures, through the connection between different platforms (i.e. grasshopper, dynamo for Revit) and eventually re-usable for several case studies.

Acknowledgment Although the contribution was conceived jointly, Massimiliano lo Turco is author of paragraphs 2, 3.1 and 5; Michele Calvano of paragraphs 1 and 4.2; Elisabetta C. Giovannini of paragraph 3.2 and Andrea Tomalini is author of paragraph 4.1.

References

Bianchini C, Inglese C, Ippolito A (2016) The role of BIM (building information modeling) for representation and managing of built and historic artifacts. Disegnarecon 9(16):10–11

Bianchini C, Nicastro S (2018) La definizione del level of reliability: un contributo alla trasparenza dei processi di Historic-BIM. Dn-Building Inf Model Data Semant 2:45–49

Brusaporci S (2017) Digital innovations in architectural heritage conservation: emerging research and opportunities. https://doi.org/10.4018/978-1-5225-2434-2

Chiabrando F, Donato V, Lo Turco M, Santagati C (2018) Cultural heritage documentation, analysis and management using building information modelling: state of the art and perspectives. In: Ottaviano E, Pelliccio A, Gattulli V (eds), mechatronics for cultural heritage and civil engineering, pp 181–202. https://doi.org/10.1007/978-3-319-68646-2_8

Chiabrando F, Farina S, Galleano A, Lo Turco M, Naretto M (2017) Photogrammetry and HBIM for the knowledge and conservation of villa sarmatoris in salmour, piedmont. In: Empler T (ed), 3D modeling and BIM. Progettazione, design, proposte per la ricostruzione, DEI Tipografia del Genio Civile, Rome, pp 230–245

Dore C, Murphy M (2012) Integration of historic building information modeling (HBIM) and 3D GIS for recording and managing cultural heritage sites. In: 2012 18th International conference on virtual systems and multimedia, pp 369–376. https://doi.org/10.1109/VSMM.2012.6365947

Fai S, Rafeiro J (2014) Establishing an appropriate level of detail (LoD) for a building information model (BIM)-west block, parliament hill, Ottawa, Canada. ISPRS Annals 2(5):123–130. https://doi.org/10.5194/isprsannals-II-5-123-2014

Giovannini EC (2017) VRIM workflow: semantic H-BIM objects using parametric geometries. In: Empler T (ed), 3D modeling and BIM. Progettazione, design, proposte per la ricostruzione, Rome: DEI Tipografia del Genio Civile, pp 212–229

Ippoliti E, Albisinni P (2016) Virtual museums, communication and/Is representation. Disegnarecon 9(17):1–15

Maiezza P (2019) As-built reliability in architectural HBIM modeling. ISPRS annals of the photogrammetry, remote sensing and spatial information sciences, 42(2/W9):461–466. https://doi.org/10.5194/isprs-archives-XLII-2-W9-461-2019

Pace S, Chiorino C, Rosso M (2005) Italia '61. Identità e miti nelle celebrazioni per il centenario dell'Unità d'Italia. Turin: Umberto Allemandi and C

Paris L, Wahbeh W (2016) Survey and representation of the parametric geometries in HBIM. 9(June)

Quattrini R, Battini C, Mammoli R (2018) HBIM to VR. Semantic awareness and data enrichment interoperbility for parametric libraries of historical architecture. https://doi.org/10.5194/isprs-archives-XLII-2-937-2018

Santagati C, Lo Turco M (2016) From structure from motion to historical building information modeling: populating a semantic-aware library of architectural elements. J Electron Imaging 26(1):011007. https://doi.org/10.1117/1.JEI.26.1.011007

Tomalini A, Lo Turco M (2019) Parametric Heritage. The digital reconstruction of P.L. Nervi's Palazzio del Lavoro in the 3D modelling Age. In: International archives of the photogrammetry, remote sensing and spatial information sciences—ISPRS archives, vol 42, no (2/W15), pp 1149–1155. https://doi.org/10.5194/isprs-archives-XLII-2-W15-1149-2019

Veron E, Levasseur M (1983). Ethnographie d'une exposition: l'espace, le corps et les sens. Centre Georges Pompidou, Bibliothèque publique d'information

Volk R, Stengel J, Schultmann F (2014) Building information modeling (BIM) for existing buildings—literature review and future needs. Autom Constr 38:109–127

Zancanaro M, Kuflikì T, Boger Z, Goren-Bar D, Goldwasser D (2007) Analyzing museum visitors' behavior patterns massimo. User modeling 4511:238–246. https://doi.org/10.1007/978-3-540-73078-1

Heritage Education for Primary Age Through an Immersive Serious Game

A. Luigini and A. Basso

Abstract Heritage education is an activity that is increasingly present in the educational paths of schools and museums. Cognitive mechanisms and representative devices must be thoroughly analysed and designed in order to promote the creation of didactic paths to foster effective learning experiences. Immersive visualization technologies are well suited for gamification applications and the technological and economic accessibility of VR HMD (head-mounted display) viewers makes these technologies particularly attractive for the development of potentially more widespread methodologies. The present contribution will describe an educational path, and its related experimentation, on the cultural heritage about the production of the typical bread of the Val Pusteria area—and the rural life around it. The project was aimed at primary school children and was based on a serious game in Virtual Immersive Reality.

1 Introduction

In 2018, around 55 million people visited the Mibac museums,[1] an increase of about 10% as compared to the previous year. As Mibac Museums General Director Antonio Lampis stated, "*Few areas are growing by 5% these days. [....] I always remember that in museums the real result to aim for is not to sell tickets, but to be able to offer, as the rules say, real experiences of knowledge*". The goal was to offer effective experiences of knowledge to visitors to museums, exhibitions and any other area in

[1]Data available on the website of the Ministry of Cultural Heritage and Cultural Activties, at the following link: http://www.beniculturali.it/mibac/multimedia/MiBAC/documents/1550245833146_2018_Musei_Tavola6_al_13-02-19.xls.

A. Luigini (✉) · A. Basso
Faculty of Education, Free University of Bozen, Bolzano, Italy
e-mail: alessandro.luigini@unibz.it

A. Basso
e-mail: alessandro.basso@unibz.it

C. Bolognesi and D. Villa (eds.), *From Building Information Modelling to Mixed Reality*, Springer Tracts in Civil Engineering,
https://doi.org/10.1007/978-3-030-49278-6_10

Fig. 1 Equirectangular panoramic views and perspectives of some interiors related to the serious game levels. Rendering software used, Octane Render by Otoy

which you can be exposed to cultural heritage. Not entertainment, not only the cult of the past, but experiences that leave an indelible mark on visitors. This is the goal that we wanted to translate from the museum environment to bring it to schools.

How can an effective experience of knowledge be built to pursue such goal? In order to suggest a methodology, it would be necessary to refer to theories of a general nature and then to decline them in the case of our interest (Fig. 1).

In the psychological field, Jerome Bruner (1956) reminded us that our thought is continuously poised between a logical-scientific dimension and a narrative dimension, where the latter refers to a deep, primordial need of our civilization to build meaning for the manifestations observable throughout life, or, in short, to interpret reality. If *logical-scientific* thinking provides clarity and organizes knowledge, narrative thinking investigates the polysemic value of knowledge and the experience of knowledge. In this indeterminate certainty, it allows individuals to find their own role within the narrative. About fifty years after the formulation of the first theories on narrative thinking, Bruner further deepened his studies underlining how two dimensions of narrative thinking are fundamental: the first is the "narrative creation of the self" as a fundamental component of the construction of a subjective dimension of identity, and the second is the interpretative dimension, which allows us to overcome the limits of given solutions. (Bruner 2002). Many other studies have developed individual aspects or entire theories of narrative thought, especially in the field of psycho-pedagogical sciences. Later literature (Egan 1986; Toolan 2001; Norris et al. 2005; Avraamidou and Osborne 2009) on the structuring of narratives useful for educational processes, have inspired some choices in the design of the Serious Game in the present work.

Among other theorizations, the seven components to build a story, proposed by Norris et al. (2005)—Event-tokens, Narrator, *Narrative appetite*, Past time, Structure, Agency, Purpose and Reader—seem to highlight important factors to be taken into account in the design of an educational path. In particular, the concept of *narrative appetite* (Norris et al. 2005 p. 541), that is, the ability of a narrative to feed the desire to discover what will happen and how the story will end, is indeed really important when creating a very concrete experience of knowledge, as the one build in our proposal.

In an essay on the cornerstone of digital techno-culture, Giuseppe Longo (1998) offers us a gnoseological reflection, underlining the diametrically opposed nature between two profiles of knowledge construction that our civilization has developed over time: the first, of an archaic profile, in which "knowledge [is] *tacit*, global and immediate, implemented by the body and embodied in its structure and its biological functions [...] guided by the affective and emotional system", and a second one, "more recent from an evolutionary point of view, [...] *explicit* knowledge, implemented in the forms of abstract logic and in general in rationality. (Longo 1998, p. 58). This new counterpoint between the emotional dimension and the rational dimension confirms how the dichotomy of the elaboration of our experiences at the psychological level necessarily requires a planning of museum experiences, or more generally educational, with an equally dichotomous approach (Luigini 2019b).

It is evident how is important, in this process of development that we are moving towards the realization of an effective cognitive experience, to focus on the user all these design cares that we are anticipating: in fact, we have applied a User Centred Design methodology.

2 Serious_Game@Brixen

The use of digital technologies for art and heritage education is a key research topic (Champion 2016; John et al. 2017; Luigini and Panciroli 2018; Challenor and Ma 2019, Luigini 2019a; Hu et al. 2019). In particular, the use of serious games seems to be increasingly widespread in the context of Digital Heritage (Ioannides et al. 2017; Basso 2019).

The VAR.HEE. project on Virtual and Augmented Reality for Heritage Education in school and museum Experience, funded by the ——, is developed in the period 2018–2021 and aims to use advanced representation and visualization technologies for the design and implementation of educational heritage education courses dedicated to both primary schools and exhibition contexts, such as museums or temporary exhibitions, and experienced in an immersive and participatory environment. One of the most important criteria that we have given ourselves was the widespread replicability of the paths, so we decided to use technologies available and commercially accessible (three-dimensional modelers and low profile visualization software and

Fig. 2 Pano2VR interface for interactive virtual tour editing. Attic level scenario

digital stereoscope like Oculus Rift) to an audience averagely literate from the point of view of computer technology (Fig. 2).

The project is developed by an interdisciplinary team coordinated by architect and digital representation scholar and in which other disciplines participate: cognitive psychology and pedagogy from the ——, as well as the partnership of scholars from the universities Alma Mater Studiorum University of Bologna, University of Camerino and University of L'Aquila.

The project during the first year was developed with regard to the design of the educational paths. The didactical path was completed from January to April 2019, while the assessment was executed from 28th to 30th of May of the same year. The didactic paths have been developed together with the teachers who have joined the experimentation. To maximize the pedagogical value of the didactic path, it has been preferred to build the activities starting from the didactic planning already foreseen for the school year 2018–19. The choice, agreed with teachers and school managers, fell on a didactic path on the water cycle and its sustainable use, which needed a complementary design to allow integration with a process of heritage education.

3 Bread and Its Tradition as a Cultural Heritage

The role of cultural heritage in primary education is increasingly important (Branchesi 2006; van Boxtel et al. 2016). Both in the academic field and in the daily practice of teachers, cultural heritage enters every day into the educational process at primary age and often transforms its overall dynamics.

After the raccomandation No. R (98) 5 of the Committee of Ministers to Member States concerning Heritage Education of 1998, which underlines the role of heritage education, it was in 2005 that the Faro Convention of the Council of Europe Framework Convention on the Value of Cultural Heritage for Society, to which Italy adheres only in 2013, which acknowledges the universal value of heritage education. In 2015, the General Direction for Education and Research of the MIBACT Ministry of Cultural Heritage and Activities and Tourism issues the fundamental Circular n.27/2018 DG-ER Piano Nazionale per l'Educazione al Patrimonio Culturale 2018–2019.

The cultural heritage speaks of our history and the civilization that has produced the present society, and for this reason it is a macro-theme that potentially affects every area of knowledge.

In the project described here, the cultural heritage is represented by all the natural raw materials, the knowledge that allows us to produce and transform them, the traditional culture that revolves around the traditional bread of the Puster Valley, such as, for example, the traditional farms and the tools of rural life. In this project, as it should be in any heritage education project, there is no rhetoric to regret the "time that no longer exists", but the awareness that handing down cultural heritage is a way to build identities, to know our origins and think about our future.

The didactic activity started with a visit to the Südtiroler Landesmuseum für Volkskunde in Teodone (BZ), where the children followed a path already organized by the museum staff, on the rural tradition and the preparation of the bread in Val Pusteria. The visits of the two classes participating in the experimentation, was before the testing of our educational paths in VR of at least 2 months.

4 Look, Move, Learn

From the point of view of methodological approach, the implementation of the serious game has required in-depth reflection and the search for innovative solutions from the procedural point of view but accessible from the technological and economic side.

First of all, we must consider that the target group were children of the 2nd and 4th years of primary school (about 7–8 years the first and 9–10 years the second), with their own cognitive structure, their own time of adaptation and their own collaborative mode. In order to adapt the device to the children, the theory of cognitive load, which explains some mechanisms capable of facilitating learning, has been taken into account from the earliest stages. In particular, the *redundancy* of textual information was avoided, capable of becoming dispersive beyond a certain threshold, favouring visual information; the whole serious game was set up with a *sequencing* process and the "*tasks*" required of the child were broken down and then reaggregated into small groups (*chunking*) and presented in modular units with the same internal sequential structure.

As a result of these evaluations, it was decided to organize several observation points, in other words "visual stations", for each environment: the first series dedicated to the observation of the environment itself, without further stimuli, and only the last following the activation of the "play" button—for the solution of the puzzles. This sequencing of the internal structure—and its decomposition—made it possible for the children-players to maintain their concentration when necessary and to be free to explore the visual space, when possible.

The Serious Game is developed in seven successive environments, which lead children to visit a farm from the outside and from the fields of wheat, through the kitchen, the bedroom, the living room (*Stube*), the barn and the roof, and then come back to the outside at the outdoor oven.

Each environment has at least two points of view: the first, dedicated to observation, and the second, dedicated to the gameplay. The reasons for such a distinction lie in the need to reduce the potential distraction generated by the need to explore during the game: in digital environments—immersive but also non-immersive—each user is naturally led to explore the space (by motor or visual means) to understand its characteristics and to build relationships between their body and the environment. Thus, the first point of view—dedicated only to visual exploration—becomes the moment when the user enters into contact with the digital environment and, for subsequent thresholds of detail, pays attention to ever more detailed information. In this way, probably, the activity to be carried out in the second point of view—the proper game—can enjoy more attention and "represent" more consistently the knowledge and skills of the player.

Once you have reached the point of view containing the game, the latter needs to be activated by a hidden button—under the feet—and visible only in the case of the passage of the pointer. This attention was necessary to avoid unintentional activations that in beta testing were numerous.

When the game was activated, the players had to demonstrate knowledge and skills acquired during the visit to the Folksmuseum in Teodone by means of quizzes or by means of predetermined activities (such as the correct selection of ingredients needed to knead rye bread).

The transition to the next environment is determined by the correctness of the quiz solution. Thus, constituting the phase of the dichotomous structure of Norris et al. (2005, p. 542).

In every environment the players were able to visually explore the rural life, in analogy with what happened during the visit to the Museum of Teodone: environments, materials, furniture, furnishings and landscape have been reconstructed through a philological study concerning the heritage of the South Tyrolean traditional farms.

The conclusion of the game is, as already mentioned, an external scenario, in which the teacher of the class—previously taken up with the technique of the green screen and inserted in post-production in the HDR 360—which summarizes some of the contents of the path, acting as a "pedagogical condenser" of the path itself: this expedient is fundamental to close the construction of the narrative path, to satisfy the narrative appetite enunciated by Norris et al. (2005, p. 541).

Particular attention was required by the vision system, both in relation to maintaining the balance of the cognitive load and for the limitation or cancellation of the effects of kinetosis or cybersickness (Al Zayer et al. 2020; Rossi and Olivieri 2019; Basso et al. 2019). More and more cases are observed of people who, during or after the use of digital stereoscopic viewers, show symptoms such as nausea, dizziness, headaches, increased sweating. Psychological and physiological sciences have been studying the phenomenon for a long time. (Reason and Brand 1975; Treisman 1977; Riccio and Stoffregen 1991), also because the onset of these symptoms is also observable in other situations, not only if subjected to a digital stereoscopic vision, such as during travel by car, train and plane or for astronauts during space flights.

Especially for children aged 2–12 years, who seem to be the subjects who most easily present these symptoms (Reason and Brand 1975), attention to the smallest detail is important to reduce or eliminate the manifestation of these symptoms.

The three main theories that have tried to explain the onset of motion sickness, progenitor of the more specific cyber sickness, are *the theory of sensory conflicts* (Reason and Brand 1975), the *Treisman's evolutionary theory* (1977) and *the theory of postural instability* (Riccio and Stoffregen 1991).

In a nutshell, the theory of sensory conflicts focuses on the discrepancy that occurs in the ocularvestibular system when our visual system and our balance system provide contrasting stimuli, causing discomfort that manifests itself with the symptoms described above.

Treisman's evolutionary theory explains motion sickness as a disturbance caused by mobility systems that conflict with the evolutionary parable of our species, which probably would need a slower adaptation to transport systems such as the car, the train or visualization systems such as digital binoculars. Another possibility that is explained in evolutionary terms is that one of the first symptoms of the intake of poisonous substances is sensory alteration, and that therefore nausea is a mechanism of selfdefence of the organism that feels attacked by a "poison".

The *theory of postural instability*, however, tells us that our organism is programmed to maintain the stability of its posture in relation to the environment in which it is located, and in the case of VR the change—sudden or not—of the surrounding environment, may produce cybersickness.

The design responses that we have elaborated foresee the reduction or cancellation of the risks listed above through some specific settings of the representation device. First of all, we have chosen to limit the interaction of the player to the visual system, choosing to realize the serious game starting from 360° static images—rendered starting from digital models—and using a *teleportation locomotion system* similar to what happens, for example, in the Google Street view system, probably already known and therefore "familiar" to the recipients of the project. Walking-based systems would probably have required greater *body involvement* of children, and in the case of locomotion managed by touchpad controllers, an increased risk of discrepancy between visual and motor stimuli.

To move from one point to another, activate game phases and interact with the quizzes, it was therefore preferred to adopt a "point and click" system that can be activated both with a pointer sensitive to the movements of the VR viewer and with the touchpad controller (but only for students of the 4th year). This choice has allowed users to reduce adaptation times to an extremely low threshold and has allowed them to enter the game in an almost natural way.

The approach is not unprecedented (Argyriou et al. 2017) but the plus-value of this work is the continuity of the research path that starts from the pedagogical project, continues with the design of the game and the architectural environments, then with the engineering and the constitution of the device of representation and then ends with the experimentation and the collection—with the subsequent analysis—of experimental data.

5 Construction Phases to Support an Interactive Reconfiguration of Spaces for the Serious Games

In the case study in question, the operations of three-dimensional reconfiguration focus on the diffusion of cultural heritage through graphic-digital models derived from reconstructive hypotheses and documentary investigations on some historical structures characteristic of South Tyrol, the Masi, pursuing the project of editing a serious game (SG) that would stimulate a young user to interest in topics related to history and local culture.

This new method of cultural dissemination is part of the wider field of Visual Tecnoculture (Gigante 1993) and Interactive Design, where the theories on *Digital Cloning and Digital Crafting for Virtual Migration* (Basso 2016) are explored as tools of contemporary communication, hypothesizing different stages of development in support of a digital reconfiguration of spaces aimed at an experience of virtual reality for education through emotional-cognitive perception (Argelaguet et al. 2016).

These operations require technical and methodological complexities that cannot be separated from digital survey methodologies, obtained through integrated methods, such as photo modelling or data acquisition through the laser scanner. In the procedures of digitization of real spaces, the study of historical data, photographic documentation and previous manual surveys carried out on the physical and cultural context contribute to develop an overall vision aimed at defining physical, plastic and chromatic characteristics that can enrich the experience of knowledge of the object detected. In addition to self-modelling processes, manual modelling can be used to create props[2] in order to improve the exploratory-immersive experience within interactive platforms compatible with HMD (head-mounted display) optical displays, where excellent care is required for the visual details (McLellan 2001).

In relation to execution time and budget, the methodological and operational procedure has therefore provided for the experimental use of different modelling software, Polygonal and NURBS, and compositing tools, useful to semantically divide the digital elements for transfer to interactive exploration platforms, together

[2] Objects—3d scene models designed to recreate the atmosphere of digital explorable environments.

Fig. 3 Automatic retopology, sculptural detailing and texturing added within Pixologic Zbrush

with rendering tools and mesh optimization tools aimed at generating visual detail during the immersive experience.

We choose to follow a workflow aimed at obtaining 3D models suitable for virtual transfer to RTR Unreal Engine 4.24 platform and at the same time suitable for the use of these assets 3d on other compositing software such as Cinema4d+Otoy Octane Render with the aim of generating 2d equirectangular images compatible with Pano2VR to simulate three-dimensional environments for editing interactive virtual tours VR[3] adaptable with active optical viewers—in the case study was used the solid Oculus Rift system.

In this regard, it was essential to follow specific measures for the optimization of three-dimensional assets in order to make the 3d elements involved in the project light in terms of digital weight without sacrificing the apparent formal complexity and therefore the visual quality of the same.

The executive pipeline includes a series of procedures that bring this particular methodology closer to that used in the editing of new generation video games, where graphics and lightness are essential to make the 3D assets exploitable on interactive platforms such as Unreal Engine, born as a platform for the creation of video games and now used in cultural operations and scientific design programs (Fig. 3).

6 Digital Cloning Procedures

The reconfiguration of complex environments has assumed, in order to obtain a realistic metric formal restitution, the simultaneous use of two reconstructive approaches: a first method related to a correct definition of the spatiality and proportions between the various compositional elements of the architectures, regarding the perception of the magnitudes of the volumes of the scene—these characteristics can be drawn

[3]The system provides for the experimental use of multiple platforms linked together within an active exploration in virtual space: is developing the ability to switch from a gaming mode from Unreal Egnine space to a virtual tour 360 exploration aimed at educational insights.

from an in-depth analysis of three-dimensional data obtainable only through procedures of digital survey—and a second method of manual modeling aimed at the specification of the various 3d assets based on the study of drawings, videos and historical photographs, with the aim of obtaining an aspect linked to the "sentimental" perception of the reconfigured environments, involving the use of the chromatic component, the morphology of the small objects and the identifying plastic details, always keeping in mind the structure of the game, with any paths, collision factors and combinations with the elements involved in the VR experience (Gershenfeld 2012; Steuer 1992).

The case study then developed two synchronic paths for the digital survey:

- Photo modelling from photo data set acquired in real sample environments (some typical farms were chosen)
- Manual Modelling from interpretation of historical images of indoor environments.

Agisoft Metashape and Meshroom are the photomodelling software used for the reconstruction of the point cloud and the mesh model that has returned the real environments surveyed to scale. A single Maso model was not used, but 3d digital information was used to understand the exact functional division of the environments and how the rooms were structured and the arrangement of the furniture. From these data, new typological models were then drawn, capable of tracing the same characteristics.

On a practical level, the workflow adopted was based on the semantic decomposition of elements by similarity of the material or through other criteria of subdivision. The simple procedure allowed a series of uninterrupted steps: the digitized models were developed in particular using Rhinoceros 3d and M.O.I 3d, as far as the modeling of macro structural elements related to the architectures is concerned, while for the props, i.e. furniture and furnishing assets, it was preferred to use some polygonal modeling functions of Maxon's Cinema 4d, software able to offer alternative tools for the creation and correction of the meshes.

C4d offers many advantages aimed at a correct transfer to other interactive platforms, such as TwinMotion, recently acquired by the EpicGames team, or Unreal Engine. It is a good idea not to change the spatial coordinates setting of the pivot associated with the models so as not to compromise the phase of exchange between the various programs used, despite today Unreal has simplified the procedure of sharing 3d resources through the use of the DataSmith plug-in. A possible phase of UV mapping and projection of the hd photographic textures on the model is delegated to Pixologic Zbrush, a famous digital sculpting software used in cinema and graphics. This software is able to manage millions of polygons at the same time and offers many useful tools to improve detail but it is essential for the automatic retopology of the exportable model (Fig. 4).

The topology obtained through autogenerative triangulation algorithms, during the transition from NURBS modeling programs to polygonal composers (export/meshing phase), can lead to the creation of an excessive number of polygons, in terms of weight byte, to be compatible with an RTR of the type of Unreal,

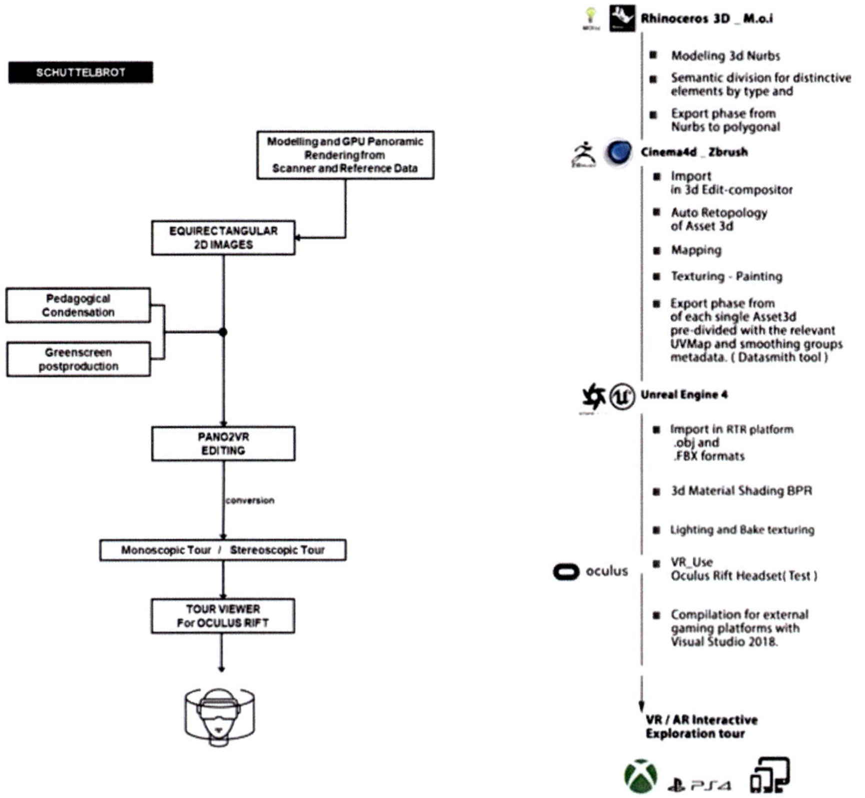

Fig. 4 Graphic scheme synthesis of the serious game development workflow

also generating multiple visual artifacts in the calculation of light maps and in the correct attribution of normal polygons.

An automatic retopology procedure (plug-in Z-Remesher) is therefore necessary, a usually very complex phase that can be completed with good results through different automatic methods within the Pixologic software, exploiting functions related to the management of Voxel algorithms and polygonal subdivision that have allowed a balanced optimization of 3D assets, texturing and the relative creation of UV Map corresponding to the new mesh generated, always distinguishing the structural and solid parts from the different decorative parts, and merging the assets associated to the material shaders with the same characteristics. To generate detail in order to make every 3d asset introduced in the serious game scenario more credible we have chosen to use Pixologic Zbrush again.

The software has recently introduced in its latest release some specific tools to add sculptural detail in a direct and flexible way using external 3d models or previous sculpting interventions made on the 3d object. The HistoryRecall brush offers the possibility to capture the depth position of carved details of a model using the Undo

History timeline, and then project those details onto a completely different model using brush strokes. The very useful thing is that HistoryRecall does not require models to have the same topology or the same number of vertices. It serves to redesign the details of the sculpture between models and to project the details of the sculpture by hand rather than through the Project All options, which is the methodology still in use today to project the detail between 3d objects. HistoryRecall allows you to take the position of a model in 3D space and store it in the Undo History slider.

After the position is stored, you can take the *History Recall brush* and ap-ply the stored status to another mesh, projecting the details to the other model. If both models do not match and overlap in the same 3D space, the projection will be faulty. This feature requires both model surfaces to be matching and overlapping in 3D space for a successful projection.

Once the 3d objects are obtained with the new topological mesh quod, it is possible to increase the levels of subdivision in non-destructive way, thus maintaining the lower polygonal levels. This feature will allow us to generate depth maps, from the normal map to the Bump maps, Displacement or Vector Displacement maps, a ploy to simulate a hd detail on a low-density polygonal model that works well in interactive software. Returning to Cinema4D, the compositing program, through a bridge system (GoZ), prepares the model for the next export phase to Unreal Engine, distinguishing the hierarchical order of the elements according to semantic criteria, defining the assets to be considered in the project as instance cloning object and building the draft of a game level then to be completed on the real time rendering platform. Cinema4d also allowed to generate 360HDR Panoramics in Equirectangular format, using the advanced Octane Render rendering engine that implements the fast Nvidia GPU unbiased calculation to offer an excellent photorealistic quality of the lighting.

The Virtual tour, compiled on Pano3VR using 360 panoramas generated with Octane, is an alternative simplified game path for unfamiliar users to a full immersive exploration of the digitized environments. The simplified Serious Game has used equirectangular images to simulate environments that can be explored in 3d from a single point of view across the various gaming sessions, the playing structure is based on simple multiple-choice questions and cognitive mini-games in which specific elements of the scenario must be selected in order to proceed with the tour. Right or wrong answers are identified through invisible buttons that follow linear input-output systems (Virvou and Katsionis 2008).

Depending on the answers, seven scenarios are explored plus the final scenario in which it is possible to acquire, in a simple and fast way, a lot of information on the local cultural heritage, on the traditional rural life and notions about typical places in South Tyrol, such as the Masi.

7 Game Design Procedures and Interactive Structuring

Unreal Engine 4.24, used in the complex development of the levels, has been selected for its excellent quality of interactive real-time visualization and for the simplicity of managing the algorithms of animation and management of input-output actions through the "Blueprint" system. Blueprints is the visual scripting system within Unreal Engine 4 and is the fastest way to develop a prototype game. Instead of having to write code line by line, you can configure input and output actions through a visual interface: drag and drop nodes, set their properties in a GUI (Graphical User Interface) and drag node links to connect. This design method is a complete scripting system based on the concept of using node editors to create game elements within Unreal Editor. As with many common scripting languages, it is used to define classes or object-oriented objects (OO) in the engine. When using UE4, you often find that objects defined using Blueprint methods are in the common language called only "Blueprint".

This system is extremely flexible and powerful as it provides designers with the ability to use virtually the entire range of concepts and tools generally available only to programmers. In addition, the specific Blueprint markup available in the C++ implementation of Unreal Engine allows programmers to create basic systems that can be extended by designers to other compatible platforms.

The software is able to easily offer an excellent fluidity of the exploration not secondary aspect of the experience because in the VR simulation it is appropriate to have a frame rate ranging from 30 fps up to 60/80 fps, to avoid disorientation and cyber sickness.

The management of simulated lighting effects and the physically correct rendering of materials has been done directly within Unreal: the impartial BPR shaders not only use Normal and Bump maps to return to the optimized model the detail of the original high-resolution polygonal model, but correctly simulate the behavior of real materials, such as translucency or opacity, while the effects of light and shadow are imprinted directly on the textures, through a procedure called Texture Baking/Lightmap. This method allows to economize the calculation of the gpu related to the simulation of global illumination made only once and automatically imprinted on the texture maps, when compiling the environment can be explored without excessive effort by the graphics card even on platforms with less performing computing power.

With the introduction of the new Nvidia RTX graphics cards it is expected that Light map methodologies will become obsolete because Path Tracing calculations will be done in real time in the future. Today this procedure is already possible but not yet applicable in scenes of great structural complexity, because the denoise system that eliminates light artifacts noise and the visual update on screen during photon simulation is still too slow and not accurate for indoor scenes. In the case study we have therefore relied on the solid method of texture baking.

For both versions of the game the implementation of the Oculus Rift, on VR Tour-viewer support, has allowed a good synchronization with the movements of the head and a correct positioning of the optical cone, thanks to its sophisticated gyroscopes and motion sensors, allowing a good virtual identification (Metz 2015). Joining the Unreal/Tour-viewer+Oculus system is a fundamental resource in terms of communicability of visual information, storytelling but also, at a macro level, in terms of the focus of interest for the conservation and enhancement of the Heritage (Bennett

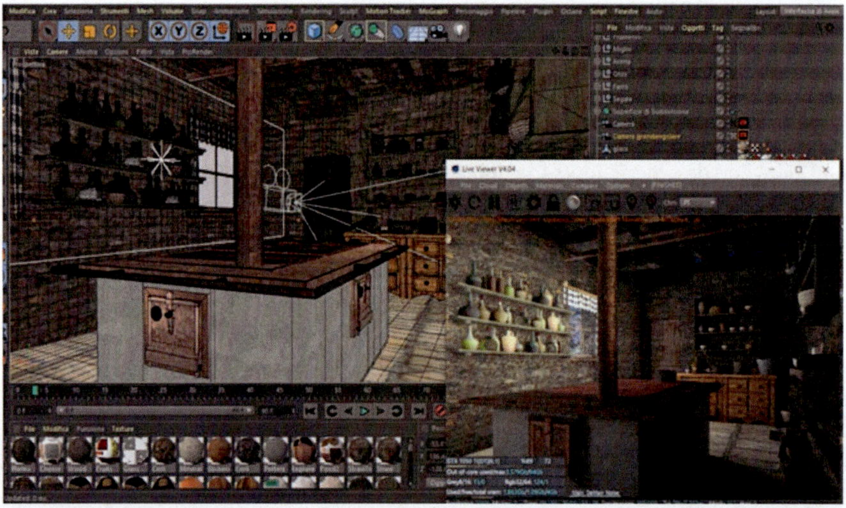

Fig. 5 Cinema4d interface as level compositor, external rendering engine used Octane Render by Otoy for photorealistic light output

et al. 2008). The choice to edit a serious game in three-dimensional environment compatible with the VR experience, is the repository of a complex conceptual value, reasonably aimed at implementing a reference system, ideal in the assumptions, concrete in the solutions and holistic in the objectives, which is a valid alternative to the canonical teaching systems (Schuemie 1999) (Fig. 5).

8 Methods

8.1 Participants

The sample included 15 students attending the second year (7-8-year-old) and 21 students attending the fourth year (9-10-year-old) of the primary school "——", located in ———. They were all checked, before participation, in order to exclude severe diseases. All of them had had normal or corrected-to-normal vision. Informed consent for their participation was obtained from their parents and from the Principal of their School.

8.2 Materials and Procedure

The reduction of adaptation times and the naturalness in engaging the game, as described in the previous paragraphs also in relation to the system of participation in the game, was very effective: the participation, in fact, was organized into several groups of 7 children, one of which was called "*operator*"—who wore the VR viewer—and the other 6 were called "*observers*"—who participated watching the projection and movements of the operator in the virtual environment on a large screen—.

Each environment was explored and activated by the operator, while the observers could interact at certain times and could help the operator in case of uncertainties or difficulties in the quizzes. At the end of the experience in one section of environment, the operator was replaced by another child in the group and joined the observers. The observations were structured in different forms: the first, during the whole experimentation, was a series of video footage with four cameras, two of which were arranged frontally and laterally to the operator's position, one was arranged to frame the observer's players and one at 360° to allow a synchronous relationship between the operator, the observers and what happened during the game in the virtual space. At the same time, a member of the research team recorded each event that was considered significant by the coordinators, so that the audio-video data could be more accurately analyzed.

Subsequently, three questionnaires were administered to the players. The first one was based on the standardized test sITQ_PQ for measuring the "presence" in virtual environments, proposed by Witmer and Singer (1998). It included 5 questions, evaluated through 5-levels Likert scales, and it was administered as soon as the operator phase was completed and before entering the *observer phase*. The second one was a test of approval of the experimentation, including other 5 rating questions on 5-levels Likert scale, which was administered at the end of the game by all members of the group. The third one, was aimed to assess the permanence of knowledge and skills acquired during the experiment and was administered at school after about a week from the VR experience and included a mixture of Likert-scale, closed and open questions, to investigate general evaluation of the VR experience and what has been learnt due to the VR experience.

9 Results

The reduction of adaptation times and the naturalness of engaging the game, also in relation to the system of participation in the game, was very effective: the participation, in fact, was organized into different groups of 7 children, one of which was called operator—who wore the VR viewer—and the other 6 were called observers—who participated by watching the projection of the operator's movements on a large screen. Each environment was explored and activated by the operator, while the observers could interact at certain times and help the operator in case of uncertainties or difficulties in the quizzes. At the end of the experience in one environment the operator was replaced by another child in the group and joined the observers. In this way the experience was shared, no child felt uncomfortable feeling alone in a virtual space for a long time[4] and in fact—even judging by the results of the questionnaires after the experiment—they experienced the game more continuously.

The observations have been structured in different forms: the first, during the whole experimentation, is a series of video footage with four cameras, two of which are arranged frontally and laterally to the operator's position, one is arranged to frame the observer's players and one at 360° to allow a synchronous relationship between the operator's behavior, those of the observers and what was happening in the game. At the same time, a member of the research team recorded each event that was considered significant by the coordinators, so as to be able to analyse the audio-video data more accurately and directly access a specific "event". Subsequently, 3 questionnaires were submitted to the players: the first, based on the standardized test sITQ_PQ for measuring the "*presence*" in virtual environments, proposed by Witemer and Singer (1998) administered as soon as the operator phase was completed and before entering the observer phase, the second as a test of approval of the experimentation administered at the end of the game by all members of the group, and the third, to evaluate the permanent knowledge and skills acquired during the experiment, submitted at school after about a week.

In summary, the results of the analysis of the data collected—which are still under processing—allow us to argue that the level of adaptation and operation of children during the experiments, thanks to the design measures mentioned above, was excellent: 76.8% respond positively or very positively to questions about adaptation in immersive environments, 67.9% positively or very positively about the confidence with immersive environments, and 94.6% positively or very positively about the visual involvement of immersive environments.

The understanding of the value of the natural landscape and the correlations between the responsible use of water, also in the domestic environment or in daily actions, and the effects of these attentions on the global climate has been good and observable in most cases.

The acknowledgement of cultural heritage as a virtuous vision of humankind's relationship with water was satisfactory in many cases—more than 70% of the tests were fully satisfactory—but not in all cases: probably the playful aspect and the satisfaction of having passed a level of the game by responding positively, may have reduced the attention of some participants by limiting the potential scope of learning. These cases are undergoing further analysis in order to identify possible improvements in the gaming device and its presentation as a function of individual diversity.

[4]The exploration and solution of each level of the game required a time ranging between 6 and 10 min which, even in the absence of experimental data on it, can be considered an effective time frame to eliminate any risk of cyber sickness due to overexposure of children to VR.

In general, the experimentation has been very positive, showing how a careful design of educational paths to heritage education—natural and cultural—through immersive serious games, is an effective methodology to build an effective experience of knowledge.

References

I Piano Nazionale per l'Educazione al patrimonio culturale (2015–2016), available at the link: http://dger.beniculturali.it/wp-content/uploads/2019/05/Primo-piano-nazionale-educazione_2015-2016.pdf

II Piano Nazionale per l'Educazione al patrimonio culturale (2016–2017) available at the link: http://dger.beniculturali.it/wp-content/uploads/2019/05/Il-secondo-piano-nazionale-d_educazione-al-patrimonio-culturale-2016-2017.pdf

III Piano Nazionale per l'Educazione al patrimonio culturale (2018–2019), available at the link: http://dger.beniculturali.it/wp-content/uploads/2019/05/PNE_allegato.pdf

Argelaguet F, Trico M, Hoyet L, Lecuyer A (2016) The role of interaction in virtual embodiment: effects of the virtual hand representation. In: Proceeding of IEEE virtual reality conference. Greenville, South Carolina, USA, pp 3–10

Al Zayer M, MacNeilage P, Folmer E (2020) Virtual locomotion: a survey. IEEE Trans Vis Comput Graph 26(6):2315–2334. https://doi.org/10.1109/TVCG.2018.2887379

Argyriou L, Economou D, Bouki V (2017) 360-degree interactive video application for cultural heritage education. In: 3rd Annual international conference of the immersive learning research network. Coimbra, Portugal 29 Jun–26 Jul 2017 Verlag der Technischen Universität Graz. https://doi.org/10.3217/978-3-85125-530-0

Avraamidou L, Osborne J (2009) The role of narrative in communicating science. Int J Sci Edu 31(12):1683–1707. https://doi.org/10.1080/09500690802380695

Basso A (2016) Il Riuso virtuale degli edifici, rivivere il Reale con lo strumento digitale. In: Reuso Pavia 2016_Contributi per la documentazione, conservazione e recupero del patrimonio architettonico e per la tutela paesaggistica. Atti del convegno internazionale, a cura di S. Parrinello, D. Besana, EDIFIR edizioni, Firenze 2016, pp 182–187. ISBN 978-88-7970-8016-6

Basso A (2019) Serious game and 3D modelling for the immersive experience development in digital tale. In: Parrinello S (ed) Digital and documentation. Pavia University Press, Databases and Models for the enhancement of Heritage, Pavia

Basso D, Saracini C, Palladino P, Cottini M (2019) Travelling salesperson in an immersive Virtual Environment: Experimental evaluation of Tracking system device. Adv Intell Syst Comput 919:519–529

Branchesi L (ed) (2006) Il patrimonio Culturale e la sua pedagogia per l'Europa. Armando Editore

Bennett AL (2008) The potential and uniqueness of virtual environments for education. New horizons in adult education and human resource development 22(3):53–59

Bruner J (1956) A study of thinking, New York: Wiley, trad. it. (1969) Il pensiero. Strategie e categorie, Roma: Armando

Bruner J (2002) Making Stories. Law, literature, life, New York: farrar, strauss and giroux, trad. it. (2006) La fabbrica delle storie. Diritto, letteratura, vita, Roma-Bari: Editori Laterza

Challenor J, Ma M (2019) A review of augmented reality applications for history education and heritage visualisation. Multimodal Technol Interact 3:39

Champion E (2016) Critical gaming: interactive history and virtual heritage. Routledge, New York/London

Circolare n.27/2018 DG-ER Piano Nazionale per l'Educazione al Patrimonio Culturale 2018–2019

Gigante MA(1993) Virtual reality: definitions, history and applications. Virtual Real Syst 3–14. https://doi.org/10.1016/b978-0-12-227748-1.50009-3

Gershenfeld N (2012) How to make almost anything: the digital fabrication revolution. Foreign Aff 91:43

Hu X, Ng J, Lee JH (2019) VR creation experience in cultural heritage education: a preliminary exploration. Proc Assoc Inf Sci Technol 56:422–426. https://doi.org/10.1002/pra2.42

Ioannides M, Magnenat-Thalmann N, Papagiannakis G (eds) (2017) Mixed reality and gamification for cultural heritage. Springer International Publishing, Cham.

John D, Shaw L, Cheetham P, Manley H, Stone AA, Blakeburn M, Gosling K (2017) Educational virtual reality visualisations of heritage sites. In: Eurographics workshop on graphics and cultural heritage, pp 189–193

Longo GO (1998) Il Nuovo Golem. Come il computer cambia la nostra cultura. Roma-Bari: Editori Laterza.

Luigini A (ed) (2019a) Proceedings of the 1st international and interdisciplinary conference on digital environments for education, arts and heritage earth 2018, Springer, Cham

Luigini A (2019b) Digital bodies. Digital experience. Digital heritage. In: Parrinello S, digital and documentation. Databases and models for the enhancement of Heritage, Pavia University Press, Pavia

Luigini A, Panciroli C (eds) (2018) Ambienti digitali per l'educazione all'arte e al patrimonio. Milano, Franco Angeli

McLellan H (2001) Virtual reality. McLellan Wyatt Digital, New York

Metz R (2015) Inside the first VR theme park. Report MIT technology review. https://www.technologyreview.com/s/544096/inside-the-first-vr-theme-park/. Access 18 May 2017

Norris SP, Guilbert SM, Smith ML, Hakimelahi S, Phillips LM (2005) A theoretical framework for narrative explanation in science. Sci Ed 89(4):535–563. https://doi.org/10.1002/sce.20063

Raccomandation No. R (98) 5 of the committee of ministers to member states concerning heritage education del 1998

Reason JT, Brand JJ (1975) Motion sickness. Academic press, New York

Riccio Gary E, Stoffregen Thomas A (1991) An ecological theory of motion sickness and postural instability. Ecol Psychol 3(3):195–240

Rossi D, Olivieri A (2019) First person shot: interactive dynamic perspective in immersive virtual environments. In: Belardi P (ed) Reflections the art of drawing/ the drawing of art, Atti del 41 convegno dei docenti della rappresentazione, Perugia, sept 2019. Gangemi, Roma

Schuemie MJ, van der Mast CAPG (1999) Presence: interacting in virtual reality. In: Proceedings, twenty workshop on language technology, vol 15, pp 213–217

Steuer J (1992) Defining virtual reality: dimensions determining telepresence. J Commun 42(4):73–93

Toolan MJ (2001) Narrative: a critical linguistic introduction. Routledge, London

Treisman M (1977) Motion sickness: an evolutionary hypothesis. Science 29:493–495

Van Boxtel C, Grever M, Klein S (eds) (2016) Sensitive pasts. questioning heritage in education (New York/Oxford: Berghahn, 2016, 294 pp., ill., isbn 978 1 78533 304 0 hardback, isbn 978 1 78533 305 7 ebook)

Virvou M, Katsionis G (2008) On the usability and likeability of virtual reality games for education: the case of VR-ENGAGE. Comput Edu 50(1):154–178

Witmer, Bob G, Singer, Michael J, (1998) Measuring presence in virtual environments: a presence questionnaire. Presence 7(3):225–240

Egan K (1986) Teaching as storytelling: an alternative approach to teaching and curriculum in the elementary school. London: Althouse Press.

M.I.R.A. Morandini, Architecture and Immersive Reality: A Realized Example of Shifting Immersive Reality from Representation to the Final Outcome

Gabriele Pitacco, Antonio Giacomin, Monica Bidoli, Marco Gnesda, Ayu Anggia Lestari, and Laura Pecar

Abstract M.I.R.A. Morandini is the acronym for Augmented Reality Immersive Museum dedicated to the intellectual and poet Luciano Morandini. The M.I.R.A. Morandini is the first prototype of an immersive museum with both real contents and virtual ones, where the visitors can move freely. M.I.R.A. Morandini shifts the use of XR from a tool for visualization and/or representation to the final outcome. M.I.R.A. Morandini is a concrete example of how architects should be thinking about XR as not just a tool to design and visualize physical spaces, but as the final product, a space to design, with attention to the spatial experience.

Keywords Virtual reality · Augmented reality · Architecture · Virtual architecture · Augmented architecture · Immersive · Museum

1 Introduction

The most interesting feature in immersive technologies research is to properly define the various aspect of "R". According to Milgram and Kishino (1994) in their "Virtuality continuum" the digital elements added to the real environment enhance the reality with the final result of a totally digital environment called VR—Virtual Reality.

G. Pitacco (✉)
Gabriele Pitacco Architetto, via Timeus 12, Trieste 1-34100, Italy
e-mail: pitacco@gp-a.it

A. Giacomin · M. Bidoli · M. Gnesda · A. A. Lestari · L. Pecar
Fluido.it, via Monte San Giovanni 23/a, Muggia, Trieste 1-34115, Italy

C. Bolognesi and D. Villa (eds.), *From Building Information Modelling to Mixed Reality*, Springer Tracts in Civil Engineering,
https://doi.org/10.1007/978-3-030-49278-6_11

Fig. 1 M.I.R.A. Morandini—the perspective section cuts through the void to show the full perceived volume of the virtual museum © Gabriele Pitacco and Monica Bidoli

The steps between reality and virtuality are known as Mixed Reality (MR) that includes Augmented Reality (AR), in which virtual digital objects are added to a real environment and Augmented Virtuality (AV), in which real objects are added to virtual ones. The immersive technologies were further defined as XR (cross reality) as "fusion of ubiquitous sensor/actuator networks and shared online virtual worlds" (Lifton and Paradiso 2009). Virtual Reality (VR), as a term, was popularized by Lanier (1988) to describe experiences where the user is entirely immersed into a three-dimensional digital virtual world and interacts with it (Fig. 1).

A literature survey in the VR definition, presented at the beginning of almost each research paper, shows, within the architecture and architecture—related fields, some kind of homogeneity in the definition of the concept, without identifying a shared one or a prevailing one.

In 1996 Stuart identifies five features describing the characteristics of VR systems that are: (1) three-dimensional viewing, (2) dynamic display, (3) users are active navigators, (4) the image displayed in the VR is from the point of view of the user's head, and (5) multimedia interaction.

Freina and Ott interestingly underline the existence of these two different kinds of VR based on viewing methods: non-immersive and immersive. The former is defined as "a computer-based environment that can simulate places in the real or imagined worlds; the latter takes the idea even further by giving the perception of being physically present in the non-physical world" (Freina and Ott 2015). The project analyzed within this document is developed within the latter. This chapter presents an example of local based immersive experience where virtual and real object are included in an experience where the user has the perception of being physically present in the non-physical world.

2 Architecture and Virtual Reality

According to the literature there are several cases where AR, VR, and XR is used in different fields either not related to architecture in any way or where architecture doesn't play any role, being considered as a neutral background for other more interesting elements. According to Paranandi et al. (2002) "although AR, VR and XR have found good applications in Medicine, Flight Simulation, and Video Game Industry, its effect on architecture remains imperceptible", but also (Wojciechowski et al. 2004; Freina and Ott 2015) and many others share a similar point of view.

If we focus on architecture and design, the role of AR, VR and XR is mainly restricted to be a (common) medium for digital representation, visualization and/or reconstruction. Digital visualization can range from the (1) reconstruction of old or destroyed building and artefacts, to (2) the prefiguration of future ones, to (3) the creation of impossible spaces and objects or to (4) other forms, like training tools.

Visualization and representation of old artifacts and objects or simulation of historical buildings can be seen in several literature survey that explore the reconstruction of lost or never built projects like, for instance a Roman Palace (Dima et al. 2014), virtual restoration of partially or fully damaged buildings and structures on historic sites to enable visitors to see them integrated into their real environment (Panou et al. 2018), visualization of incomplete or broken real objects as they were in their original state by superimposing their missing parts (Liarokapis and White 2005) or even museum objects, generic 3D gallery designed for presenting cultural objects in a virtual room, a virtual exhibition presenting museum artifacts in a 3D room being a reconstruction of a real gallery—an exhibition corridor in the Victoria and Albert Museum in London (Wojciechowski et al. 2004).

As Roussou perfectly underlined "the use of architectural detail in immersive real-time virtual reality systems is difficult due to the technical and performance restrictions placed by the realtime image generator". From an architectural disciplinary perspective this can be considered a critical issues, especially in art and restoration, where the attention to the "materiality" and detail is crucial and the artwork is already a form of representation given by the artist or the architect (Fig. 2).

The most explored topic in literature is the visualization and representation of designed future entities being them buildings, artworks, objects or other things. According to Schnabel (2001) the manipulation of virtual environments during the design process pushes designers to better perceive space, for example its fluidity and functionality, without using 2D representations. VR gives the possibility to experience sensations and movement in an artificial environment that is a simulation of some aspects of the real world. According to Milovanovic, focusing on the field of architecture, VR applications' utilization are wide, from design itself, construction and project's communication as well as collaborative decision-making (Milovanovic et al. 2017).

In all the aforementioned cases AR, VR, and XR have been used as design medium, components or representation tools, just like a pencil, a design table or a CAD

Fig. 2 Marina Morandini (daughter of the poet Luciano Morandini) in front of a Getullio Alviani artwork, has been the very first visitor of the M.I.R.A. Morandini. © Luca D'Agostino (Phocus Agency)

software, while their potential as a final outcome, as a design subject instead of a design tool, is yet to be explored and exploited.

In 1969 Superstudio turned their final outcome of a design process into strongly political drawings. The final outcome of the design was not intended to be a building or a monument but the negative utopia represented (Lang and Menning 2003). The medium turned into the final outcome of the design process. That's the conceptual inspiration for this design of the project presented here.

3 Spatiality in (*Virtual*) Reality

According to Brett (2016) "Architects are trained and experienced in spatial design—spatial thinking, traditions of spatial theory, a language of spatial communication and representation, and an archive of precedent and typology are some of the skills that uniquely qualify architects to contribute to the spatial design of virtual reality experiences. Virtual reality experiences designed without attention to spatiality misses the most important opportunity of the technology." This works show an example of how architects should be thinking about AR, VR, and XR as not just a tool to design and visualize physical spaces, but as something to design in and of itself, with attention to the spatial experience.

Virtual reality can either replicate the physical world and its rules or imagine a completely different one, detached from reality. Virtual reality space offers different opportunities and shows different limitations than physical space. Light, physics, navigation, materiality, construction, and environmental issues are different in the virtual world, and the solutions and design choices require a different approach able to incorporate and explore the possibilities of their brand new spatiality.

The virtual reality also presents its own limits and constraints. A new virtual economy based more on performances than on budget, an economy deeply based on the calculation speed of the used hardware. Optimization is the new budget—polygon counts, complex lighting calculations, high numbers of textures—these all impact processing speeds that can affect battery life, hardware requirements, and frame rates.

The choice between a physical replica, a brand new world or all the in-between options must be deeply considered. According to Brett "There is value in the familiar. Skeuomorphic design takes advantage of archetypal physical forms to present something recognizable to the user. While virtual space design should not necessarily replicate all the details of physical space, designers of virtual reality spaces can reference familiar forms and organizational structures that will help users understand and navigate virtual space. Certain physics may no longer impose the same limitations on form and movement in virtual reality space as it does in physical space, but designers should not necessarily see this as an imperative to throw away all conventions. Forms and materials may no longer have the same functionality in virtual reality space, but they may still express certain behaviours and qualities through reference to their physical analogues. Where physical constraints no longer dictate form and space, designers may look to psychological factors in usability—affordance, signifiers, and conceptual models—for guiding forces in virtual space design" (Brett 2016).

4 A Museum Exhibition of AR, VR, and XR

In the past years museum have seen the exploitation of multimedia techniques and lately the introduction of virtual reality methods to create new forms of presentation for exhibitions. Virtual Reality can offer a number of advantages to museums, offering a way to overcome some common problems like the lack of space or the need of visitors to interact with the exhibits. According to Lepouras, a broad categorisation of virtual museums reveals that they vary from fully immersive cave systems to simple multimedia presentations. (Lepouras et al. 2001). Several examples can be found from the XR Guide for the Guggenheim Museum of contemporary art in Bilbao (Abawi et al. 2004), to the Manchester Jewish Museum (Tom et al. 2016), from the tourist guide for cultural heritage sites located in the old town of Chania, Crete, Greece (Panou et al. 2018), to the British Museum (Rae and Edwards 2016), the Great North Museum (Atkinson 2015) and also the Grant Museum, DiMoDA (Digital Museum of Digital Art), immersion room at Cooper Hewitt Smithsonian Design Museum, Met Cloisters in New York City, Hermitage, MAAT, the new museum of Art, Architecture and Technology, in Lisbon, Muse in Trento, (Symbola 2019), but none of these offered the kind of experience presented here.

5 M.I.R.A. Morandini. Augmented Reality Immersive Museum "Luciano Morandini"

The M.I.R.A. Morandini (Augmented Reality Immersive Museum dedicated to Luciano Morandini) was presented on Saturday 7 September 2019, on the 10th anniversary of the passing of the intellectual and poet Luciano Morandini, in his home-town, San Giorgio di Nogaro (Udine), Italy.

The M.I.R.A. Morandini is the first prototype of an immersive museum with both real contents, such as a selection of paintings from the Morandini Legacy (which collects the works of some of the post-war Friulian masters like Zigaina, Spacal, Alviani, Bassi), and virtual ones, like volumetric holograms of the sculptures, 360° immersive videos of performances, multimedia works dedicated to the poet and the reading of the poems by Luciano Morandini. The Museum prototype, tested here for the first time, is partly real and partly virtually "augmented" with more contents, more space and a contemporary architecture and design. The Museum can be visited through the latest-generation virtual reality headset Oculus Quest, released in May 2019.

Born from an idea of the architect Gabriele Pitacco (www.gp-a.it), the project was created with the XR project concept and development by Antonio Giacomin (www.fluido.it), the support of the association S@NGIORGIO 2020 for the exhibition contents and BIC Incubatori FVG srl—center of excellence on incubation and business development as certified regional incubator. The Scientific and Technological Research Area of Trieste—Area Science Park was involved for the technical and operational coordination and the Department of Architecture and Civil Engineering of the University of Trieste for the research of perceptive and spatial aspects. Comunicarte curated the graphic design, while the Municipality of San Giorgio di Nogaro acted as a host structure.

The prototype is part of the project "Hacking Real Space" funded by the Friuli Venezia Giulia Region on the Call of the European Fund POR FESR FVG 2014-2020, Action 2.1b.2 "*Grant for the financing of personalized acceleration and company consolidation programs, aimed at the entrepreneurial development of projects of cultural value, aimed at cultural and creative companies*".

6 A Museum in a Room (*Il Museo, Cielo Incluso, in Una Stanza*)

In the House of Poetry dedicated to Luciano Morandini, in a dark room looking empty (Fig. 3), a virtual reality headset is the passe-partout for accessing a unique museum: the space multiplies and expands to accommodate the world of the poet, iridescent and ever-changing, and the passage from one scenario to another is a journey in which poetry is the means of transport and the reader/visitor the protagonist.

The Morandini Augmented Reality Interactive Museum is a unique experience, in which the visitor is immersed in a world that does not follow the same rules as the real one, where existing physical elements and virtual elements interpenetrate and complement each other. Once worn the virtual reality headset, the actual space multiplies in height, generating further floors which divide, move away from each other and expand into the virtual space to create an exciting central space.

It is a space defined by a triple height void (Figs. 4 and 5) that visually connects the different thematic areas. A void that allows the museum to be perceived in its entirety and able to provide orientation and points of reference to the visitor who enters a space different from the real one enriching the visit of unexpected and exciting points of view and gives the space a unitary vision.

Fig. 3 The physical entrance to the M.I.R.A. Morandini in San Giorgio di Nogaro

Fig. 4 M.I.R.A. Morandini—the render of the poetry floor (level-1) of the Museum shows one of the text of the Morandini poems floating in space

The museum is designed in sections as a sequence of floors, cut by a diagonal beam of light entering from the roof. Light creates a central void that defines a series of floors staggered in space, a raumplan of spaces with different size, height and character. Spaces tailored on the works exhibited and, at the same time, visually connected to each other.

Movement between the different levels is possible through a series of virtual elevators, portals for teleportation -quoting Star Trek- which allow you to jump from one floor to another, almost immediately.

The width of the central void, while being shifted in space, is always constant: there is a fixed dimensional relationship between the various planes, obtained—in complementary pairs—by the subdivision of the real floor. The larger floors are the result of a virtual extension to the outside, while the smaller ones are obtained by the subtraction of this same surface. By placing all the floors along a central imaginary magic line it is possible to reconstruct the real space.

Fig. 5 M.I.R.A. Morandini—the render shows the view from the upper level to the lower ones, showing the visual connection between them and the void

The virtual museum is closely linked to real space and the existing physical building. The entrance to the virtual museum takes place on the first floor, as in the real museum. The visit then takes place freely descending to the other floors and returning to the first floor at the end, to exit.

The lower floor, at the end of the path, becomes the meeting point and the passage between virtual and real, where the real artworks of the Morandini Legacy are visible.

Exactly like the work of poet, the museum presents, within its unitary structure, a series of environments that are different from each other, but capable of interpenetrating, communicating and influencing each other.

7 Hold the (*Magic*) Line

To expand the limited size of the original room, a "magic line" was invented (Fig. 6). The imaginary line that separates the two halves of the real room is stretched to the thickness of the empty space. This stretched void creates a completely new virtual space, which in the virtual building corresponds to the central passage and thus separates the planes that are on one side and the other of the void. This filter, which corresponds in reality to a very narrow strip in the plan, allows to virtually generate an enormous environment that is completely permeable, through which you can look out at other planes and see what happens on them. In this way, the entire real surface of the room is exploited at its most, several times, with different proportions and contents in the different virtual levels.

The areas of the larger levels are obtained extruding the original volumes towards the outside, and although they are not walkable, they visually look like that they are, giving the idea of depth and welcoming all those works and objects that need to be observed from a distance, such as multimedia videos and blow-ups of the poet's photo-portrait.

In the area delimited by the magic line there are the virtual elevators, the portals of teleportation: a square space projecting on the void, straddling the magic line, which gives access to the following floors.

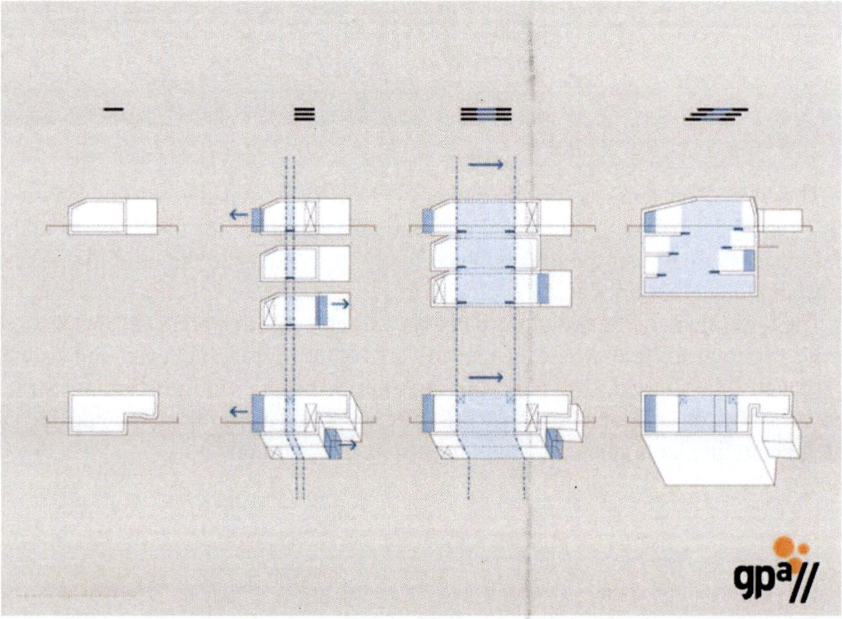

Fig. 6 M.I.R.A. Morandini—the plans (above) and section (below) diagrams shows how the magic lines expands the space creating both virtual voids (light blue), and virtual space (dark blue) that can be seen but not experienced

In the same way, the central empty space may not be walkable as it is protected by glazed parapets, but for the bravest who want to try walking on the air, there is a unique experience to be tested.

8 Images and Words

The museum opens with a biographical section (Fig. 7) and a brief note on Luciano Morandini, curated by Carlo Londero, poet's scholar, with the voice of Fabrizio Gaio recorded at the Casa della Musica (House of Music) in Trieste, together with the Luca D'Agostino's photographic portraits, curated by Phocus Agency and a 360° immersive experience. The 360° immersive experience is meta-space that bring a whole park inside a floating solid looking sphere that can be crossed.

The next level contains a reasoned selection of the poetic works of Luciano Morandini (Fig. 8), curated by Luisa Gastaldo Morandini with Carlo Londero, whose reading was made by Chiara Dorigo (Poesie a Manovella) and Rossella Gorgoglione at GLB sound of San Giorgio di Nogaro (Udine). The words of the poem are floating in space, close to walls without touching them.

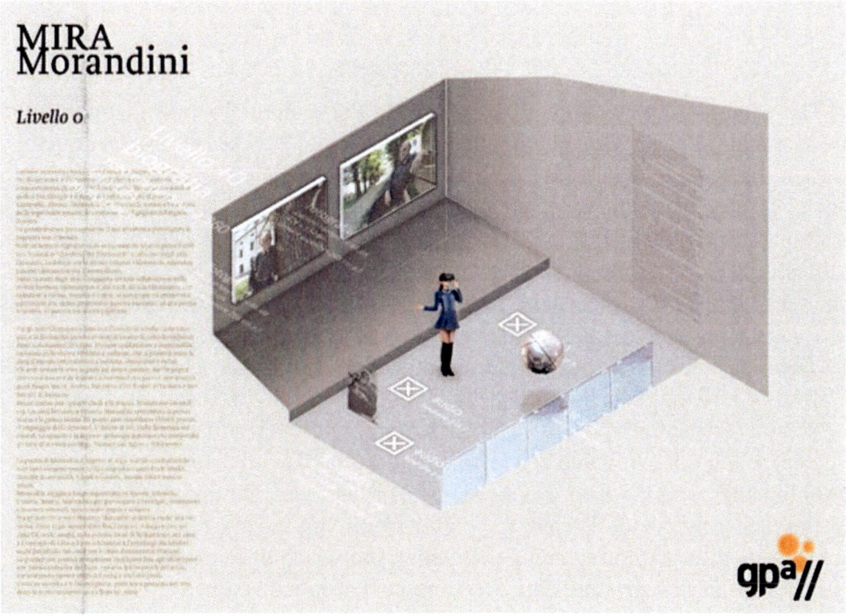

Fig. 7 M.I.R.A. Morandini—the axonometric view of level 0 shows both the floating sphere (an object that can be passed through to enter the garden of Villa Dora, a meta-space that is wider than the virtual one already perceived) and the virtual space that outdistances the two photographic portraits

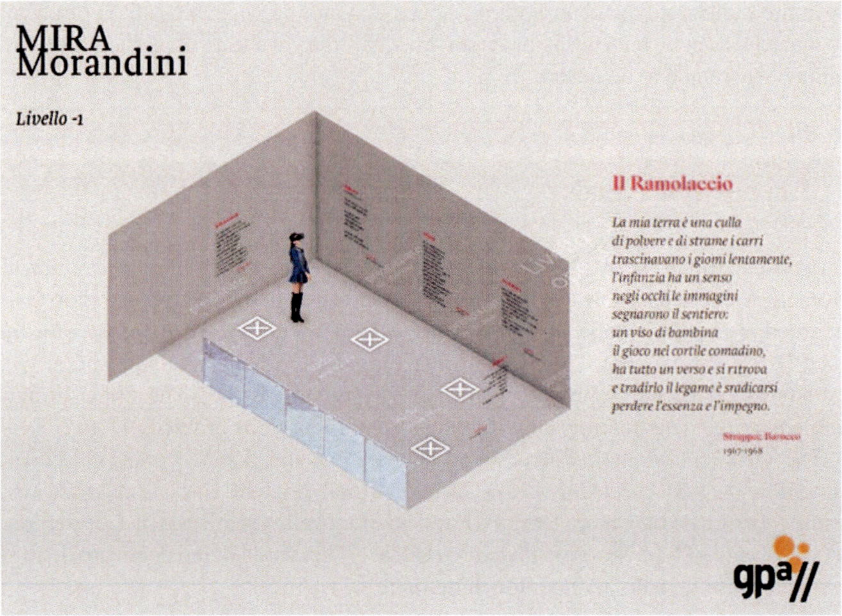

Fig. 8 M.I.R.A. Morandini—the axonometric view of level-1 shows the layout of the poetry floor, as in Fig. 4

The volumetric holograms of a selection of the sculptural works that constitute the Morandini Legacy fluctuate in the next space (Fig. 9). The selection, curated by Massimo Premuda of the Casa dell'Arte (House of Art) of Trieste, emphasizes a work by Dora Bassi, the only female artist in the collection, and an abstract work in bronze by Giovanni Piccin. The volumetric holograms have been realized with a three-dimensional photographic scan realized by Fluido.it by Antonio Giacomin. The song "Vento" performed by Elsa Martin with music by Stefano Battaglia on texts by Luciano Morandini and recorded at Swiss Radio and Television RSI—Radio 2 has been selected by Gianluca La Boria from the works presented at the International Festival of Jazz and Poetry "Festival J&P" of San Giorgio di Nogaro (Udine).

Remote Transitions (Fig. 10) is a contemporary art performance and the central element of the following floor. Created by the Cultural Association S@NGIORGIO 2020 together with the Estonian Academy of Music and Theater in Tallinn and the Tartini Conservatory of Trieste, the performance can be enjoyed in an immersive 360° video with spatialized audio that allows the spectator to experience an extract of the event in first person (Fig. 11). Remote Trasitions is a performance realized between three "poles" connected at a distance. In the first two poles (the Tallinn Music Academy and the Tartini Conservatory of Trieste) there are musicians (with traditional instruments, but also computers) connected via the Lo.La system (an acronym for Low Latency) in real time. In the third pole (San Giorgio di Nogaro) the two A/V flows of Tallinn and Trieste arrive in streaming, but with a time delay;

Fig. 9 M.I.R.A. Morandini—the axonometric view of level-2 shows the Statues floor with the tridimensional hologram of the Dora Bassi artwork

Fig. 10 Remote Transition—the photos show the synchronized performances in Tallin (left) Trieste (top-center) and San Giorgio di Nogaro (bottom-center and left)

nevertheless the double-bass player Giovanni Maier improvises on the music coming from Tallinn and Trieste, and dancer Nina Alexopoulou improvises on the music, while an interactive video by Florence-based Studio RF follows her movements in real-time.

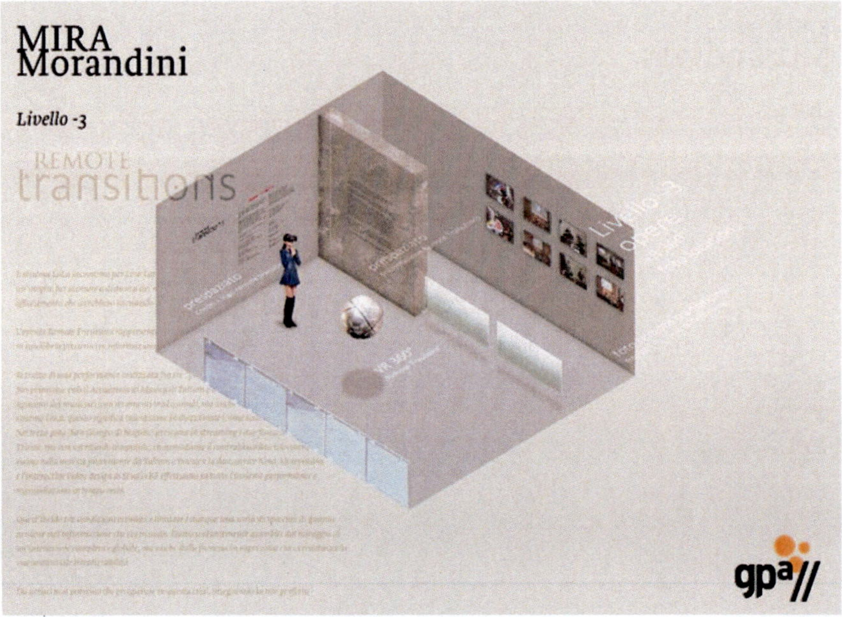

Fig. 11 M.I.R.A. Morandini, the axonometric view of level-3 shows the second meta-space dedicated to "remote transition", an immersive 360° immersive experience showing the art performance connecting in real time Trieste, Tallin and San Giorgio di Nogaro

At the end of the visit (Fig. 12) there is a selection of the graphic and pictorial works of the Morandini Legacy, with a selection of the works of some of the post-war Friulian masters, Alviani, Anzil, Chersicla, Marini, Spacal, Zigaina made available by the Municipality of San Giorgio di Nogaro. This selection is the bridge, the link between physical reality and virtual reality, connecting the two worlds though the overlapping of the artworks.

The museum (Fig. 13) represents an all-round virtual experience, able to involve all senses thanks to a varied exposition in the forms of art and in the represented themes.

9 Aspects of XR Development

Implementing a technology that interacts between real and virtual world has many issues. The main issue in a Location Based Experience (LBE) is to maintain the reference between real and virtual space. The inside-out tracking method, used by the headset, analyse in real time the environment defining the borders of the virtual experience setting the common spots in the two realities. The real environment survey

Fig. 12 M.I.R.A. Morandini—the perspective plan of level-4 shows the overlap of virtual and real. The paintings and artworks are both in the real floor and the virtual one, connecting them to create a local based immersive experience where virtual and real object are included in an experience where the user has the perception of being physically present in the non-physical world

has to be perfectly aligned with its virtual reproduction. Once the real/virtual relationship is defined the architect/designer can modify the virtual world structure to recreate and redefine the space. Another relevant issue is the interactivity between two different virtual spaces or crossing the above mentioned "magic line". The main research and development was focused on the teleportation system using a trigger system that allows to move virtually without moving physically enhancing the space experience.

In the prototype are also inserted a meta-spaces in form of a sphere where the user can enter in an environment bigger than it's container.

The whole system was prototyped and developed within a game engine (Figs. 14 and 15).

Fig. 13 M.I.R.A. Morandini—the perspective section close up shows the different floors and the exhibition lay-out

10 Conclusions

Ivan Sutherland, the creator of one of the world's first VR systems in the 1960s, stated: "The ultimate display would, of course, be a room within which the computer can control the existence of matter. A chair displayed in such a room would be good enough to sit in. Handcuffs displayed in such a room would be confining, and a bullet displayed in such a room would be fatal" (Sutherland 1965). The road to get there is still quite long but MIRA Morandini shows a concrete example and an argument for why architects can add values in the design not only of real spaces, but also for virtual ones.

Fig. 14 M.I.R.A. Morandini—the photographs show a selection of the reactions of the first 100 users that visited the M.I.R.A. Morandini

Fig. 15 Luciano Morandini, portrait © Luca D'Agostino (Phocus Agency)

References

Abawi DF, Los Arcos JL, Haller M, Hartmann W, Huhtala K, Träskbäck M (2004) A mixed reality museum guide: the challenges and its realization

Atkinson R (2015) Virtual reality experience created for great north museum: hancock. Museums Association. Last updated Apr 22 2015

Brett E (2016) Architects in the design of virtual reality spaces. UC Berkeley College of Environmental Design, Master of Architecture Thesis

Dima M, Hurcombe L, Wright M (2014) Touching the past: haptic augmented reality for museum artefacts. In: International conference on virtual, augmented and mixed reality. Springer, Berlin

Freina L, Ott M (2015) A literature review on immersive virtual reality in education: state of the art and perspectives

Hugues O, Fuchs P, Nannipieri O (2011) New augmented reality taxonomy: technologies and features of augmented environment. In: Handbook of augmented reality. Springer, Berlin

Javornik A (2016) The mainstreaming of augmented reality: a brief history. Harvard Business Review. N.p

Jerald J (2016) The VR book, human-centered design for virtual reality. In: ACM books #8, association for computing machinery and morgan & claypool publishers

Lang P, Menning W, Superstudio S (2003) Life without objects, Skra, Milan

Lanier J (1988) A vintage virtual reality interview, whole earth review

Lepouras G, Charitos D, Vassilakis C, Charisi A, Halatsi L (2001) Building a VR museum in a museum. In Richir S, Richard P, Taravel B (eds), Proceeding 2001 VRIC laval virtual international conference, Laval, France 16–18 May 2001

Liarokapis F, White M (2005) Augmented reality techniques for museum environments. Mediterr J Comput Netw 1

Lifton J, Paradiso JA (2009) Dual Reality: merging the real and virtual. FaVE, (2009) First international conference on facets of virtual environments. Steigenberger Hotel Berlin, Germany

Milgram P, Kishino F (1994) A taxonomy of mixed reality visual displays. IEICE Trans Inf Syst E77-D(12)

Milovanovic J, Moreau G, Siret D, Miguet, F (2017) Virtual and augmented reality in architectural design and education: an immersive multimodal platform to support architectural pedagogy. In: 17th International conference, CAAD futures 2017, Gülen Çağdaş, Mine Özkar, Leman F. Gül and Ethem Gürer, Istanbul, Turkey

Negroponte N (1970) The architecture machine: toward a more human environment. (MIT Press)

Panou C, Ragia L, Dimelli D, Mania K (2018) An architecture for mobile outdoors augmented reality for cultural heritage. ISPRS Int J Geo-Inf

Paranandi M, Sarawgi T (2002) Virtual reality on architecture: enabling possibilities

Rae J, Edwards L (2016) Virtual reality at the british museum: what is the value of virtual reality environments for learning by children and young people, schools, and families? MW2016: Museums and the Web 2016. Published Jan 28

Schnabel MA, Kvan T, Kruijff E, Donath D (2001) The first virtual environment design studio. In: 19th eCAADe conference proceedings, Helsinki, Finland

Stuart R (1996) The design of virtual environments. McGraw Hill, New York

Symbola Foundation (ed. Sturabotti, Surace),Museum of the Future, Mu.SA Project, 2019

Tom D, Claudia M, Jung T (2016) Value of augmented reality to enhance the visitor experience: a case study of manchester jewish museum. e-Review Tourism Res

Wojciechowski R, Walczak K, White M, Cellary W (2004) Building virtual and augmented reality museum exhibitions. In: Web3D symposium proceedings

AI-BIM Interdisciplinary Spill-Overs: Prospected Interplay of AI and BIM Development Paradigms

Lorenzo Ceccon and Daniele Villa

Abstract BIM and AI—along with other cutting-edge technologies – are converging towards a data-rich, sematic approach to modelling and design. We analyse some of the current uses of AI in combination with BIM technology, as well as some very recent approaches which may disrupt the paradigms in use within the AEC field. The most promising features suggest a huge potential for liberating the designers from the most repetitive tasks and even endow them with a whole baggage of best practices, as dynamically emerging from the available data stored in BIM models and elaborated by AI. However, in order to achieve this, some issues must be dealt with, especially as regards the knowledge sharing framework. Furthermore, the potentialities of AI in the field of architecture poses questions about the function, goals and ultimately the role of (human) designers in the world of AI-aesthetics and practice.

Keywords BIM · AI

1 Introduction

Over the last decades, the adoption of digital technologies within the field of architectural design has gradually shifted from the pure "assistance" in automatizing tasks as they were performed in the pre-digital era—as in the case of CAD—Computer Added Design—to an increasingly unknown territory, where the very same nature of the design process is getting impacted by technology and pushed towards new workflow paradigms. Some workflows have been based on the widespread use of simulations, as for the generative design tools, while others have consisted in attempts to combine several disciplines, datasets and dimensions in one unique modelling tool, which

L. Ceccon · D. Villa (✉)
Department of Architecture and Urban Studies—DAStU, Politecnico di Milano, Milan, Italy
e-mail: daniele.villa@polimi.it

L. Ceccon
e-mail: lorenzo.ceccon@polimi.it

C. Bolognesi and D. Villa (eds.), *From Building Information Modelling to Mixed Reality*, Springer Tracts in Civil Engineering,
https://doi.org/10.1007/978-3-030-49278-6_12

could be used by different persons and teams for various purposes along the whole lifecycle of a building: it is the case of the Building Information Modelling—BIM.

More specifically, BIM technology is built on a database of multidimensional information centred on the Industry Foundation Classes, a standardized set of classes and families underlying the creation of each respective object instance, e.g. a wall, a window, a roof (https://www.buildingsmart.org/standards/bsi-standards/industry-foundation-classes). In fact, BIM technology is Object Oriented, and imposes the user to think in terms of categories based on a semantic subdivision of the architecture: walls, doors, windows are each both a class—a sort of overarching category—and specific elements in the model, where specific parameters differentiate and make each instance unique, e.g. geometrical coordinates and features, materials, construction time-frame. Such an approach has many potential benefits, mostly stemming from the data-richness of the model, which is seen as a centralized mine for different purposes, ranging from representation, to planning, to estimates, to simulation, to management and many more. However, along with the huge potential of such a technology, BIM is also confronted with challenges and new possible usages which derive from the sheer quantity of the available data within a model and across models and which are handled with difficulty by even the expert users.

In this framework, the emerging Artificial Intelligence technologies seem a perfect complement to BIM, as they are already very effective at getting the most out of huge datasets, finding patterns and making useful predictions. In particular, AI is particularly promising in linking raw data from sensors—both during the site survey phase, such as laser scans, and during the construction and management phase (also as regards dwellers), especially through IoT devices—to the model and its semantic structure, potentially helping in forecasting and avoiding issues (e.g. safety during construction, energy consumption caused by unoptimized design or management), as well as integrating missing data (e.g. the hidden parts of existing constructions), and help tuning the model itself as best fitting the present and the future whole lifecycle performance. AI may even help better structure the semantic ontologies underlying BIM models (Fig. 1).

However, AI itself is far from being a mature technology—even though it has been around already for decades—and we are still in a single-task oriented, Artificial Narrow Intelligence (ANI), far from the prospected Artificial General Intelligence (AGI) and even further away from the Artificial Super Intelligence (ASI). This development stage must be considered when we try and devise the impact AI can have on BIM and the AEC sector now and in the future.

As a final remark, within the current AI developmental stage, we can already find different working paradigms, whereby a trade-off is usually present between the amplitude and reach of AI tasks and its process traceability: for instance, the most promising predictive and generative AI applications are based on Neural Networks (NN), where the intelligibility of the "reasoning" followed by the algorithm is usually low even for its developers. This aspect has great significance too, at least in view of scientific validation of choices, as well as of personal accountability for said choices (Figs. 2 and 3).

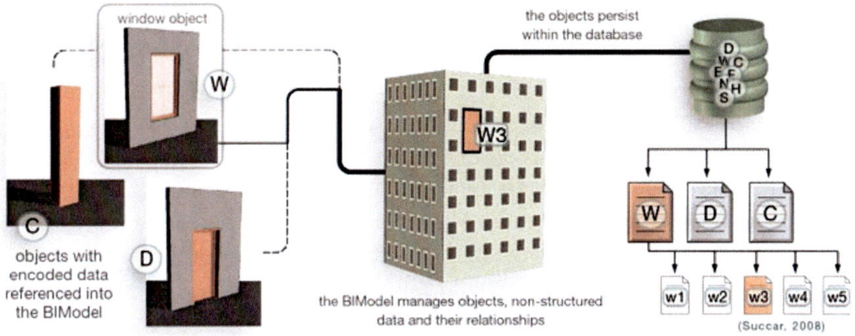

Fig. 1 "Building information models and their objects—flow diagram", Fig. 8. In building information modelling framework: a research and delivery foundation for industry stakeholders—Bilal Succar 2009

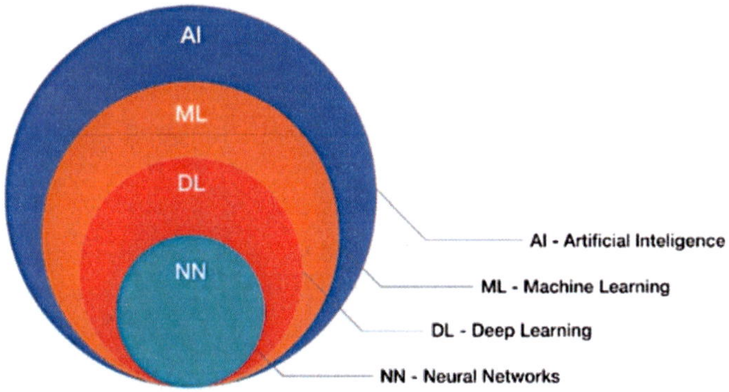

Fig. 2 Rakesh Elamaran, medium.com, Dec 30, 2018. https://medium.com/fnplus/introduction-to-machine-learning-8c3fb57a79b6

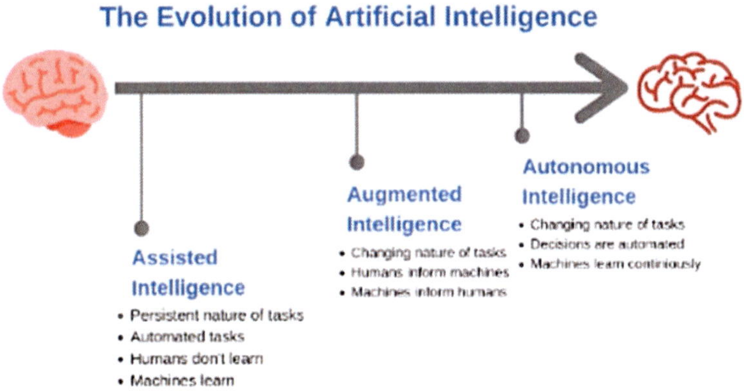

Fig. 3 https://hqsoftwarelab.com/about-us/blog/internet-of-things/bringing-ai-to-the-internet-of-things-as-if-we-aren-t-afraid-of-the-judgement-day

Technological convergence. Overall, the links between AI and BIM are still in their infancy stage, also considering that both techniques are not consolidated yet. However, it seems that AI could be the needed factor for developing a smarter BIM modelling technology, where the data explicitly or implicitly present in the model—including those stemming from sensors, IoT devices and simulations—can produce new data layers linked to the other datasets within the BIM model, thus reinforcing the best practices and devising even better ones over time through a performance-based feedback loop. The two technologies are naturally converging, since both are designed to work with a huge amount of digital data.

2 AI and BIM: Semantic Model Creation

AI has already entered in the workflow of creating BIM models, both for the existing built environment and for entirely new projects. While most of the process is common to such two cases, the case of the built environment needs a preliminary phase which is, at the same time, a challenge to the surveyors/modellers and an opportunity for the use of AI techniques.

2.1 AI and H-BIM: Survey and "Bimification"

We will then first analyse such a preliminary survey-to-BIM or "bimification" phase in "Heritage BIM" (H-BIM), and then move to the common challenges in creating a semantic-enriched model for both H-BIM and BIM. Any H-BIM project starts with the mapping the existing (built) environment, possibly not limited to the observable geometric features, but also encompassing a number of other dimensions and features, such as decay processes, materiality, hypothesized hidden structures, and many more. Over the last decades, such process has entered a new phase. While traditional survey techniques tended to smoothly and implicitly link, step-by-step, a limited number of survey data to their symbolic representation on paper—by drawing and annotating man-made measurements, observations and even hypothesized elements with the (implicitly semantic) architectural drawings—today we are facing a whole new scenario and many new challenges. In fact, new survey techniques on the one hand—such as laser scanning, photogrammetry and other digital techniques producing a data-rich flow of data over a number of possible surveyed aspects and dimensions (not just geometrical, but also chemical, thermal, …)—and the use of the (explicitly) semantic-based BIM technology requires a great effort to match these two worlds: that of raw and "dumb" data, and that of "classified" model objects and features.

Best practices in the field still tend to heavily rely on the human contribution to build the semantic model from the surveyed data. Typically, point clouds are imported in the BIM environment as a sort of "guideline" for drawing the IFC-based elements of the BIM model. Sometimes this step is performed passing through a 3D modelling package (such as Rhinoceros) for easier modelling of irregular shapes, and then the 3D model is imported into BIM packages for the semantic enrichment and for the addition of further data multidimensionality.

Over the last few years, some research projects as well as commercial packages have started to offer a sort of "assisted" solution, whereby some steps are performed automatically in order to facilitate the still needed human intervention. Two main categories have emerged, as follows.

(Point Cloud) Data Segmentation. Some solutions offer algorithms, often AI-based, to recognize semantic elements within the surveyed data, in particular within point clouds, which are "segmented" into clusters belonging to each semantic (subobject) (Figs. 4, 5, 6 and 7).

The "segmented" data can then be used as a more readable set of pre-interpreted survey dataset, through which the human task of building a semantic BIM model is made much easier and less prone to errors, also due to the easier comparability between each segmented data cluster and each semantic model element. Some of segmentation tools were added as plugins to point cloud processing software.

Recent research has been focussing on advancements on indoor mesh classification using AI, whereby different survey datasets from interior spaces are properly segmented and coordinated, in order to offer a more comprehensive help to extract the relevant spatial information (see Runceanu and Haala 2018).

Bimification. Another approach—both in research and as software package solution—directly aims at (semi)-automatizing the "bimification", i.e. "the process to

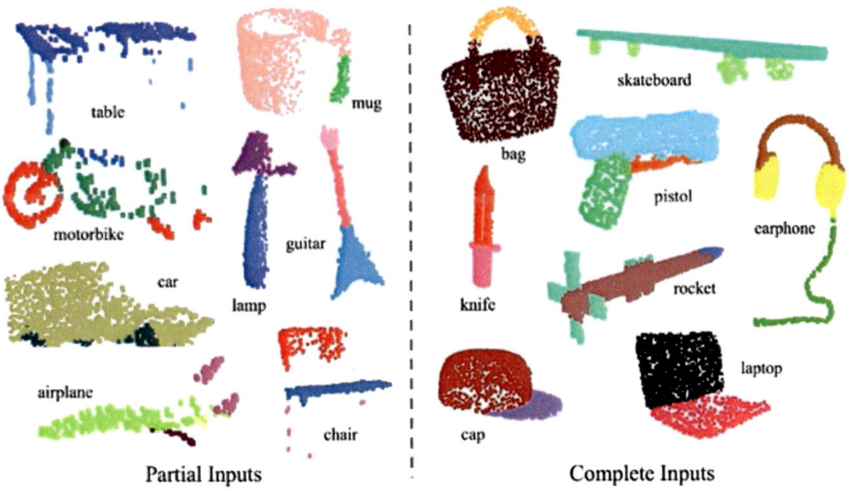

Fig. 4 "Part segmentation results. We visualize the CAD part segmentation results across all 16 object categories. We show both results for partial simulated kinect scans (left block) and complete ShapeNet CAD models (right block)", Fig. 3 in Qi et al. 2017, PointNet: Deep Learning on Point Sets for 3D Classification and Segmentation, http://stanford.edu/~rqi/pointnet/

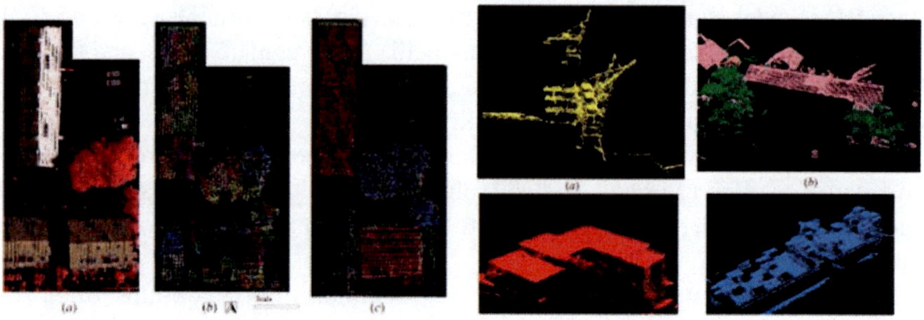

Fig. 5 Respectively, "different stages of segmentation: **a** coloured point cloud; **b** supervoxels, **c** segments" and "labelled and POINT cloud representing various objects. **a** Cranes; **b** trees and gabled roofs; **c** buildings; **d** ships". All images by Ramiya Anandakumar, A supervoxel-based spectro-spatial approach for 3D urban point cloud labelling, July 2016 International Journal of Remote Sensing 37(17):4172–420. https://doi.org/10.1080/01431161.2016.1211348

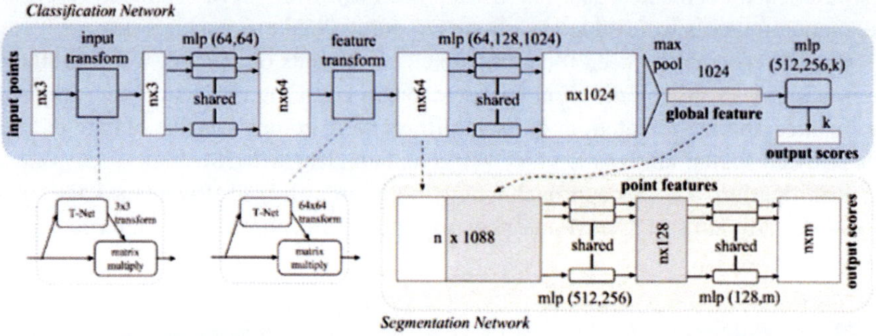

Fig. 6 Segmentation Network, http://stanford.edu/~rqi/pointnet/

obtain from an existing real facility a 3D BIM, preferably based on the standard ISO 16739 data schema (IFC), while taking into account the actual state and performance parameters of the facility" (Raimar and Scherer 2017). In other words, the semantic elements by which the survey data are sorted are directly derived from the BIM IFC classes, hence ontologies, so that an almost automatic conversion is possible. Unlike point cloud contour extraction and unsupervised clustering algorithms, where the possible output does not have to fall in any pre-defined category, here data segmentation is not performed just by clustering groups of similarly-featured data elements, rather by sorting data into pre-determined object classes. (For IFC and ontology inference, through best fit of existing ones see Rasmussen et al. 2017) (Fig. 8).

For instance, Pointfuse® claims to offer "A semi-automatic process to create simple parameterized BIM models quickly and easily. The BIM workflow accelerates the use of reality capture in space management and lease survey workflows. Walls, windows and doors within the mesh are converted to simple BIM families without the need for a library of objects. All editing takes place within Pointfuse itself. No plugins for third party software are required. Mapped to the standardized IFC schema,

Fig. 7 "PointCNN deep neural network labeled raw XYZ point cloud", Dmitry Kudinov, Jul 24, 2019. PointCNN: replacing 50,000 man hours with AI—PointCNN: replacing 50,000 man hours with AI, https://medium.com/geoai/pointcnn-replacing-50-000-man-hours-with-ai-d7397c1e7ffe

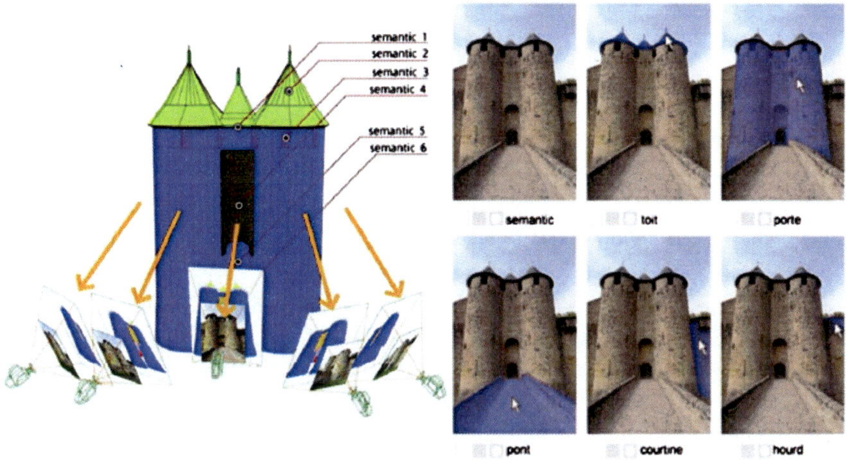

Fig. 8 Respectively, "Semantic annotation of photographs by projecting the building morphology organized according to the chosen description structuring. Segmentation of the 3D representation is mapped on oriented images structured as semantic layers" and "Interactivity of semantic layers associated to images. Moving the cursor on photos, the semantics coming from the 3Dmodel is highlighted showing vocabulary terms (on the bottom of each image) related to the selected entities", Figs. 5 and 13, images by Philippe Véron, in A semantic-based platform for the digital analysis of architectural heritage, April 2011, Computers & Graphics 35(2):227–241, April 2011, https://doi.org/10.1016/j.cag.2010.11.009

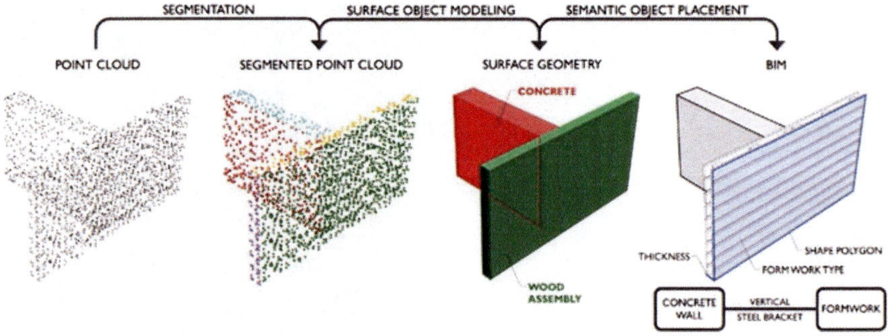

Fig. 9 "The process of generating semantically-rich SGMs from point cloud data and how material recognition can help with segmentation", Fig. 1 by A. Dimitrov, M. Golparvar-Fard, Vision-based material recognition for automated monitoring of construction progress and generating building information modeling from unordered site image collections" in Advanced Engineering Informatics 28 (2014) 37–49. https://meyar.co/wp-content/uploads/2017/02/Vision-based-material-recognition.pdf

Pointfuse sBIM exports 2D floor plans, 3D BIM models and PDF reports. This brings comprehensive reporting and easy integration to existing facility management operations and systems." (https://pointfuse.com/software/) (Fig. 9).

Artificial Intelligence is being introduced in such software packages in order to not only achieve an auto-perfecting tool, but even to make good use of the "implicit" experience related to how buildings are laid out on average. It is the case of the "Bimify" tool within Bricsys software package. At the moment it is only adding the semantic dimension of BIM model to 3D drawings: "It's a tool that uses machine learning to analyze the whole drawing and automatically add BIM data. BIMIFY detects; the classification of 3D Solids and block references, rooms, external walls, spatial locations (buildings and floors), profiles. BIMIFY also creates new custom profiles and adds them to the project's library if it doesn't find them in the standard library. It even automatically generates sections. The list goes on and on…".[1] However, it was interestingly noted that "The next challenge for Bricsys is to apply this AI to the scanning world. Working with Leica and HOK, the plan is to scan buildings internally and externally and let the AI interpolate between the external and internal meshes, to interpolate the voids where the scanners can't penetrate. It's possible to 'infer' the walls, windows, floors, doors and columns. Scan-to-BIM may actually become a reality" (Day 2019).

Nevertheless, an efficient AI workflow would require constantly closing the feedback loop between surveyed data and survey outcome with the minimum human effort. In fact "creating deep learning classifiers requires large labeled datasets; and creating large labeled datasets requires elaborate crowdsourcing systems and many hours of manual human effort applied to classification and data entry" (Czerniawski and Leite 2019). If data labelling had to be performed manually by each user or team, this would translate into extended time, and/or increased outsourcing costs,

[1] https://blog.bricsys.com/the-lasses-who-%e2%80%8eautomatically-add-bim-data/.

for training and updating the algorithms. "Fortunately, much of this effort can be bypassed in the building industry because of as-built building information models (BIMs), a semantically rich form of facility information. From these BIMs, semantics can be transferred to point clouds. This paper presents a method for creating large labeled datasets for training deep neural networks to semantically segment point clouds of buildings. Geometry and attached semantics are extracted from a BIM. The geometry is registered with the point cloud and the BIM semantics are copied to the points in the point cloud. The presented method enables organizations with access to as-built BIMs to forgo the effort of creating large labeled datasets and instead use the embodied effort in their pre-existing BIMs." (Czerniawski and Leite 2019) (Fig. 10).

If this is the cutting-edge status of research, some aspects seem still missing, especially as regards the use of further data types and data sources for an ever improved "bimification". As to data types—as we will see when dealing with the whole topic of simulation, digital twin, and backwards simulation—including data outside the domain of geometrical features (in the spectrum of the visible light) may result in a much more powerful insight on critical, yet non observable aspects of the building. Infrared techniques, decay patterns, and alike can in fact provide "clues" to better "smart-guess" what is behind the surveyed surfaces. On the other hand, data sources are of paramount importance when we are dealing with AI, where the more (quality) data, the better. In particular, while BIM IFC classes are a limited set of ontology-based object types, their real-life deployment may greatly vary. To this respect, it seems that linking AI algorithms to widely shared databases/repositories could improve exponentially the insight of AI, as we will see as a recurring Leitmotiv within this field.

As a final remark, we see already great advances about "smart-guessing" non-observable reality features in other disciplines, and we are convinced that such approaches could soon produce great spill-overs into our discipline. For instance,

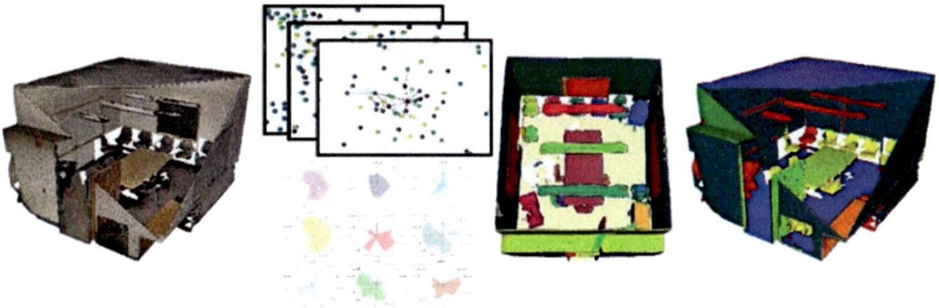

Fig. 10 "Voxel-based three-dimensional (3D) semantic segmentation. From left to right: raw point cloud, feature engineering, connected Elements extraction, Classified point cloud", by Florent Poux, Fig. 1 in Voxel-Based 3D Point Cloud Semantic Segmentation: Unsupervised Geometric and Relationship Featuring vs Deep Learning Methods, May 2019 International Journal of Geo-Information 8(5):213, https://doi.org/10.3390/ijgi8050213

the unobserved structure of tree branches has been correctly guessed and reconstructed starting from a non-complete view of the plant through "a method for inferring three-dimensional (3D) plant branch structures that are hidden under leaves from multi-view observations. Unlike previous geometric approaches that heavily rely on the visibility of the branches or use parametric branching models" this "method makes statistical inferences of branch structures in a probabilistic framework. By inferring the probability of branch existence using a Bayesian extension of image-to image translation applied to each of multi-view images, our method generates a probabilistic plant 3D model, which represents the 3D branching pattern that cannot be directly observed. Experiments demonstrate the usefulness of the proposed approach in generating convincing branch structures in comparison to prior approaches" (Isokane et al. 2018). We cannot predict how easy it would be to extend such technology to smart-guessing hidden architectural elements—tree structure has been long modelled through L-shapes and similar algorithms, unlike generic architectural elements—but the availability of copious BIM model datasets may allow to base predictions on those as well (Fig. 11).

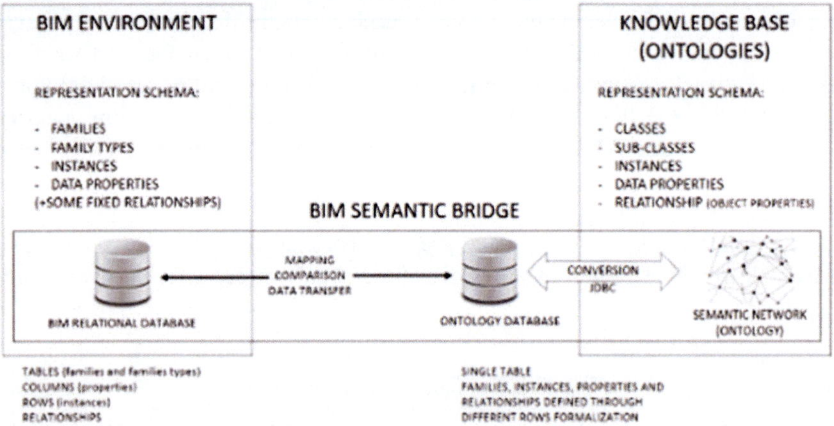

Fig. 11 "The conceptual schema of the BIM semantic bridge application", Fig. 1 in an ontology-based platform for BIM semantic enrichment, Stefano Cursi, Davide Simeone, Ugo Maria Coraglia, http://papers.cumincad.org/data/works/att/ecaade2017_050.pdf

2.2 *AI and BIM Semantic Model Creation, Based on Shared Ontologies*

BIM modelling is based on the typification of architectural object classes as agreed upon globally within the buildingSMART not-for-profit organization. "IFC is a standardized, digital description of the built asset industry. It is an open, international standard (ISO 16739-1:2018) and promotes vendor-neutral, or agnostic, and usable capabilities across a wide range of hardware devices, software platforms, and interfaces for many different use cases" (https://www.buildingsmart.org/standards/bsi-standards/industry-foundation-classes/). Object naming, classification and reciprocal relationships can be best described by the "ontologies" underpinning IFC. "An ontology is a formal, explicit specification of a shared conceptualization" (Pauwels 2019). Knowledge Graphs can be very effective to visualize them (Fig. 12).

While IFC seem to have brough a smooth and straightforward "standardized" modelling of any real-life or imagined architectural design, the choice of the appropriate ontologies is not a trivial aspect of the use of BIM technology. In fact, the

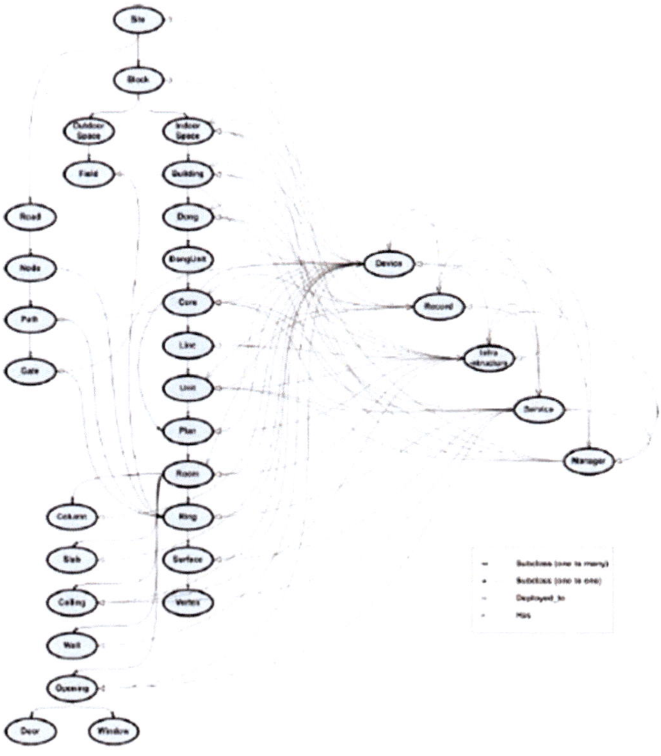

Fig. 12 "Relation of building components", Fig. 4 in building ontology to implement the BIM (building Information modeling) focused on pre-design stage, https://www.irbnet.de/daten/iconda/CIB11184.pdf

"semantic segmentation" of any building may depend both on the idiosyncratic viewpoint of the specific user, as well as on the purposes of the model itself. More specifically, since every BIM class brings a predefined set of parameters and interrelations with other classes, the choice of the best "fitting" one may be performed according to the required dependencies. For instance, whether a building uppermost envelope has to be classified as a "roof" or as a floor slab could be debatable, depending on the architecture at stake but also on the processes allowed by the BIM software package for the applied family. In fact, windows may be "allowed" to sit only on walls, while a slope could be a parameter only available for roofs by default. Semantic subdivision and BIM package default consequences must then be considered together in view of building a functional and coherent model.

Furthermore, in order to benefit from the knowledge accumulation globally, we believe that best practices about the use of standard ontologies, as well as the creation of specific ones for specific building typologies, ages and techniques, should be discussed and shared, as already experimented in some pilot projects. For instance, the INCEPTION project is as a good example of said needs to both tailor the ontology substrate of the BIM model on the specificities of the (heritage) use cases, and of the usefulness of a shared/standardized best practice thereof. "The INCEPTION project has defined the approach and the methodology for semantic organization and data management toward H-BIM modelling, and the preliminary nomenclature for semantic enrichment of heritage 3D models. The organization of consolidated knowledge is performed following a specific workflow in order to get them suitable for their reuse into H-BIM semantic model, accordingly to digital documentation and capturing protocols that have been developed" (Maietti et al. 2018) (Fig. 13).

Moreover, an even more effective use of BIM technology—especially as regards the linking of the BIM model with sources outside the traditional BIM environment, such as sensors and GIS tools—require a careful tailoring of the semantics underlying the BIM model and the translation thereof into SemanticWeb technologies. IFC INFRA and ifcOWL are, among others, widely used tools to reach these aims. "However, it is important to understand the difference between BIM standards and the Semantic Web technologies. Indeed, the ifcOWL is defined as a serialization of an IFC schema definition, in order to enable the use of contents semantically managed. For this reason, within INCEPTION, H-BIM is meant to be an ontology to support storage of semantic knowledge available for Cultural Heritage buildings and architectural complexes, as well as their related information. The integration of Getty vocabularies and ifcOWL is a complex but feasible task, focusing on the actual queries, communications and current H-BIM work. Furthermore, within INCEPTION, we have the chance to test such integration, achieving positive results as well as beginning an implementation, thanks to the use of a «glossary of names» gathered by the Demonstration Cases analysis. Once data are collected and aggregated thanks to the use of semantic technologies, the most valuable goal is strictly related to dissemination capabilities. Indeed, the richness of the INCEPTION project is represented by accessibility to a crowdsourced database of both scans and reconstructions, with different level of confidence by their source data. Therefore, the INCEPTION platform will have to contain, visualize, manage, update, and exchange technical and

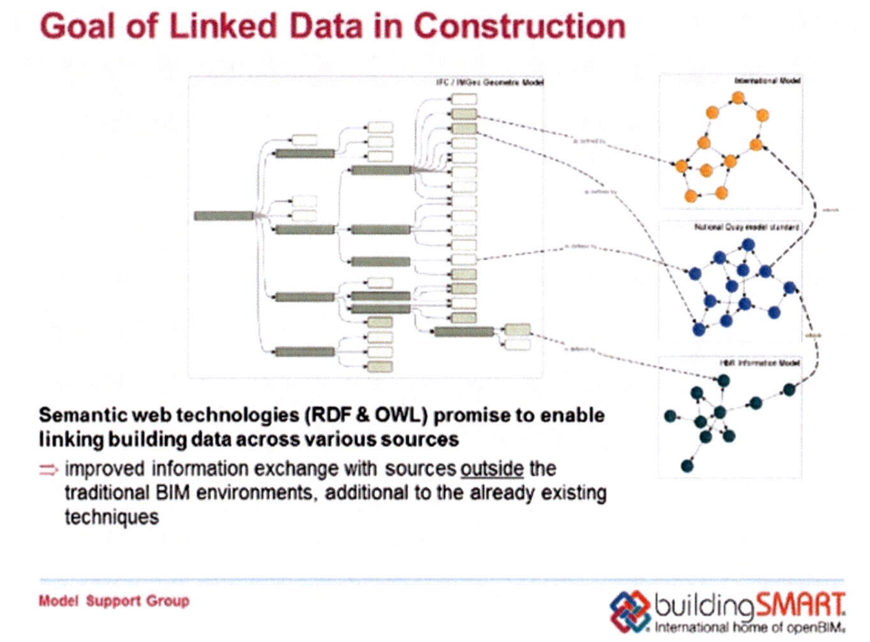

Fig. 13 "Schematic overview of the suggested principles to enrich legacy IFC models with semantically rich information captured in distributed, decentralized and easily extendible models captured in the resource description framework (RDF)", Fig. 13 in 3D spatial infrastructure for the port of rotterdam, Zlatanova et al. 2014, GISt Report No. 67 TU Delft, https://www.researchgate.net/publication/269688807_3D_Spatial_Infrastructure_for_the_Port_of_Rotterdam

divulgative information regarding historical heritage, through the use of 3D BIM models. Any form of digital or digitized content, stored or linked into a 3D H-BIM model, as well as semantically indexed by the use of the INCEPTION ontology, will allow the use of different navigation systems" (Maietti et al. 2018).

Other projects and research endeavours have explored new ways to check and reorganize ontologies based on principles of non-redundancy and modularity (see the W3C best practice rule), especially as regards domain-specific ontologies, and also trying to fill contingent gaps within them, all in view of a shared knowledge resource. Among such projects, "BIM4EEB uses the Linked Data approach as a glue between different data sources at technical and syntactic levels to make data interlinked and accessible. Adapting the Linked Data approach at the semantic level to renovation domain, through the identification and development of appropriate vocabularies is one of the goals of WP3. The first task introduces the overall Linked Data framework, outlines the vocabulary requirements in the renovation domain, analyses existing ontologies covering specific areas of the required vocabulary, and identifies gaps where new ontologies should be developed, or existing ones extended or refined. To facilitate the shared ontology work, it also specifies the tools and practices for ontology development, publication and documentation are specified. In the following tasks, documents' scope is to provide extensions of the BIM standard ontologies by properly refining relevant ontologies referring to occupants' behaviours and comfort, energy performance of systems and components, and acoustics. Also an ontology for BIM data representation on different levels of detail corresponding to renovation process modelling will be delivered. An ontology for renovation workflows will make it possible to represent the connection between activities and domain entities that can be directly observed, thus paving a way to systems that can maintain in real-time situational awareness about the renovation process. The development and harmonization of ontologies proposed by BIM4EEB partners will be discussed in CEN Technical Committee 442 to start working on a European standard on Linked Data for the AEC sector" (BIM4EEB Editorial 2019).

BOT-Building Topology Ontology, in turn, tries to devise a way to avoid contradiction between IFCs and custom ontologies, by creating a core of standard IFC-compliant ontologies and some methods to extend domain-specific ones. "In the last years, several ontologies focused on structuring domain specific information within the scope of Architecture, Engineering and Construction (AEC) have emerged. Several of these individual ontologies redefine core concepts of a building already specified in the publicly available ontology version of the ISO standardised Industry Foundation Classes (IFC) schema, thereby violating the W3C best practice rule of minimum redundancy. The voluminous IFC schema with origins in a closed world assumption is likewise violating this rule by redefining concepts about time, location, units etc. already available from other sources, and it is furthermore violating the rule of keeping ontologies simple for easy maintenance. Based on all the available ontologies, we propose a simple Building Topology Ontology (BOT) only covering the core concepts of a building, and three methods for extending this with domain specific ontologies. This approach makes it (1) possible to work with a limited set of core building classes, and (2) extend those as needed towards specific domain ontologies that are in hands of business professionals or domain-specific standardisation bodies, such as the European Telecommunications Standards Institute (ETSI), buildingSMART, the Open Geospatial Consortium (OGC), and so forth" (Rasmussen et al. 2017).

AI becomes particularly relevant in this field of semantic modelling, especially as to "Ontology refinement" and possibly even in devising new ontologies. However, given the still relatively early stages of development, the following examples are not directly linked to the BIM ontologies, but constitute nevertheless a fascinating perspective also in such domain.

A first promising project has adopted Machine Learning to check and optimize the "PrOnto" ontology, … "PrOnto was developed through an interdisciplinary approach called MeLOn (Methodology for building Legal Ontology) and it is explicitly designed in order to minimise the difficulties encountered by the legal operators during the definition of a legal ontology" (Palmirani et al. 2019). The paper presented "a refinement of PrOnto ontology using a validation test based on legal experts' annotation of privacy policies combined with an Open Knowledge Extraction algorithm. Three iterations were performed, and a final test using new privacy policies. The results are 75% of detection of concepts and relationships in the policy texts and an increase of 29% in the accuracy using the new refined version of PrOnto enriched with SKOSXL lexicon terms and definitions" (Palmirani et al. 2019).

As to the use of AI for devising new ontologies, recent attempts have been made to "develop a novel domain-independent automatic ontology generation framework that converts unstructured text corpus into domain consistent ontological form. The framework generates KGs from unstructured text corpus as well as refine and correct them to be consistent with domain ontologies. The power of the proposed automatically generated ontology is that it integrates the dynamic features of KGs and the quality features of ontologies" (Elnagar et al. 2020). As stated, the method is not aimed specifically at BIM ontologies. However, being domain-independent, it seems that it could be extended, through further research, to assist hypothesizing new frameworks for an improved, comprehensive and rationalized semantic grounding of BIM models.

Moreover, AI can play a key role in streamlining the "translation" among different artificial knowledge agents. In fact—given that "very often different ontologies are developed independently for the same domain" and that "such «parallel» ontologies raise the need for a process that will establish alignments between their entities in order to unify and extend the existing knowledge"—a research team has recently presented "a novel entity alignment method which we dub DeepAlignment. DeepAlignment refines pre-trained word vectors aiming at deriving ontological entity descriptions which are tailored to the ontology matching task. The absence of explicit information relevant to the ontology matching task during the refinement process makes DeepAlignment completely unsupervised. We empirically evaluate our method using standard ontology matching benchmarks. We present significant performance improvements over the current state-of-the-art, demonstrating the advantages that representation learning techniques bring to ontology matching" (Kolyvakis et al. 2018). Such research, currently outside the scope of BIM modelling, could find a ground-breaking application in view of matching ontologies among different BIM models and between BIM and other environments, also in view of effective data sharing repositories.

3 AI and BIM. Simulations and Digital Twin

As seen, the (H)-BIM model creation phase is starting to benefit from AI technology, and it seems that the potential is much larger. However, the analysed benefits are not entirely disruptive, in the sense that they essentially boil down to help built more reliable and more complete models in a shorter time, with less errors.

A more disruptive use of AI for the BIM industry can be seen in the use of AI as a enabling technology to link in real time the virtual and the built environment, creating a self-adjusting "Digital-Twin" of reality within the BIM world. Such an endeavour, based on the availability of real-time data from IoT devices and sensors linked to a semantic-organized multidimensional model, can benefit from the ability of AI to make predictions based on both real and simulated data.

Thus, on the one hand, AI could help fine-tune the model correspondence to reality, based on a backward simulation process (model fitting)—where real-life data are compared to the simulated outputs in view of be closely matched by the model—and, on the other hand, AI could help optimize several parameters during the whole building lifecycle, from construction to management and maintenance—including structural static conditions and building indoor comfort and energy efficiency—along the paradigm of Model Predictive Control (MPC), which promises more effective results than the more traditional Rule-Based Control (RBC) in view of Building Management System (BMS). "Many studies have proven that the building sector can significantly benefit from replacing the current practice rule-based controllers (RBC) by more advanced control strategies like model predictive control (MPC). However, the optimization-based control algorithms, like MPC, impose increasing hardware and software requirements, together with more complicated error handling capabilities required from the—commissioning staff. In recent years, several studies introduced promising remedy for these problems by using machine learning algorithms. The idea is based on devising simplified control laws learned from MPC. The main advantage of the proposed methods stems from their easy implementation even on low-level hardware. However, most of the reported studies were dealing only with problems with a limited complexity of the parametric space, and devising laws only for a single control variable, which inevitably limits their applicability to more complex building control problems. In this paper, we introduce a versatile framework for synthesis of simple, yet well-performing control strategies that mimic the behavior of optimization-based controllers, also for large scale multiple-input-multiple-output (MIMO) control problems which are common in the building sector. The approach employs multivariate regression and dimensionality reduction algorithms. Particularly, deep time delay neural networks (TDNN) and regression trees (RT) are used to derive the dependency of multiple real-valued control inputs on parameters. The complexity of the problem, as well as implementation cost, are further reduced by selecting the most significant features from the set of parameters. This reduction is based on straightforward manual selection, principal component analysis (PCA) and dynamic analysis of the building model. The approach is demonstrated on a case study employing temperature control in a six-zone building, described by a linear model with 286 states and 42 disturbances, resulting in an MPC problem with more than thousand of parameters. The results show that simplified control laws retain most of the performance of the complex MPC, while significantly decreasing the complexity and implementation cost" (Drgoňa et al. 2018) (Fig. 14).

HVAC dynamic control. One of the currently most promising field where MPC is being adopted in the building thermal control management, both as research topic and in some commercial software packages.

As to the former, "the use of Model Predictive Control (MPC) in Building Management Systems (BMS) has proven to outperform the traditional Rule-Based Controllers (RBC). These optimal controllers are able to minimize the energy use within building, by taking into account the weather forecast and occupancy profiles, while guaranteeing thermal comfort in the building. To this end, they anticipate the dynamic behaviour based on a mathematical model of the system. However, these

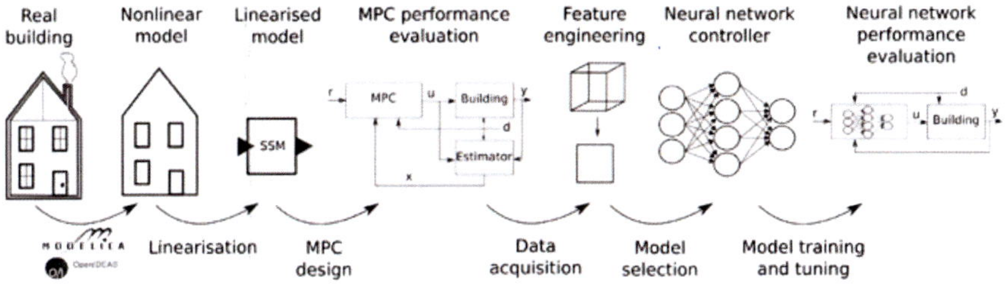

Fig. 14 "Schematic view of the methodology. From left to right: a real building is modelled using the building energy simulation (BES) Modelica library IDEAS. The obtained nonlinear model is then linearized and converted to a state space model (SSM) representation. The model is employed for both emulating the building dynamics in the simulations, as well as a controller model in the MPC. The MPC is being implemented and evaluated in a MATLAB® environment through on-line coupling. The simulation data are collected and reduced in dimensions by selecting only the most significant feature variables. Next, the machine learning model is selected and further trained and tuned to mimic the behavior of the original 'teacher' MPC. Finally, the performance of the original and approximated MPC is being evaluated and compared with traditional controllers", Fig. 1. In Drgoňa et al. 2018, p. 202, https://doi.org/10.1016/j.apenergy.2018.02.156

MPC strategies are still not widely used in practice because a substantial engineering effort is needed to identify a tailored model for each building and Heat Ventilation and Air Conditioning (HVAC) system. […] a Python-based toolbox, named Fast Simulations (FastSim), that automates the process of setting up and assessing MPC algorithms for their application in buildings, is presented. It provides a modular, extensible and scalable framework thanks to its block-based architecture. […] FastSim is developed and used in a research environment, however this automated process will also facilitate the implementation of MPC for different building systems, both in virtual and real life" (Arroyo et al. 2018).

As to the latter, a good example is provided by the software Comfy by Intel and Building Robotics. "According to Building Robotics and Intel (Intel provides IoT gateways enabling the physical connection of the Building Management System (BMS) to the cloud) the Comfy app can collect data from existing BMS in commercial spaces and also take into account user request data to moderate and optimize temperatures in different parts of the building dynamically with the aim of increasing cost savings and improving energy efficiencies in commercial spaces. Once Comfy receives a request from a user, it automatically provides 10 min of hot or cool air to gauge the discomfort levels based on how many additional requests are received. Over time, the app uses user request data to identify patterns and preferences based on location and time of day" (Intel Editorial 2015).

Damage Detection in near real-time. Predictions about the state of a building over time, especially as regards damages caused by hazardous events, such as earthquakes, require to be timely and possibly continuous in time in order to be effective. In such cases, whether it is about the static condition of a bridge or the safety of a building, what matters most is the accuracy/reliability and the prompt-availability of the prediction, so that immediate action can be taken. Researchers have been experimenting AI-based solutions in this field, which could revolutionize the concept of monitoring itself.

A first piece of research focussed on *masonry building*. "The research reports on the achievements using two digital technologies, such as Building Information Modelling (BIM) and Artificial Intelligence (AI) in the field of damage detection and assessment of masonry buildings. The purpose of the work is to point out the potentials and the benefits for the assessment and the automatic digitization of target buildings affected by hazard-induced damages. To this scope, the photogrammetry technique is used to fit with the implementation of on-site visual data into the technologies previously mentioned. The final aim is to speed up the way of approaching intervention strategies on existing buildings that are becoming obsolete in relation to the technological progress of the building sector" (Musella et al. 2019).

Another team focussed on *stone-walls*. "A large proportion of the data created during the inspection and assessment of stone facades and their damages is recorded in formats that are not machine-readable and thus cannot be further processed or managed digitally. Consequently, this increases the risk of data loss and incorrect information due to human misinterpretation. Therefore, a Multimodel-based approach has been developed in which stone facades of existing buildings are digitized as IFC-model by using proxy entities and linked with web ontologies for semantic enrichment. Additionally, detected anomalies in the stone structure are implemented and linked with geometrical representations. By utilizing additional rules and inference mechanisms, the anomalies can be classified, and a knowledge-based damage assessment is processed" (Seeaed and Hamdan 2019).

Overall, as a common phenomenon underlying the foregoing examples, technological convergence—or at least the useful spill-overs and mutual reinforcing attitude of Digital Twin and AI—has been clearly highlighted by Alanen 2019, in the following terms. "The first benefit of a digital twin is the ability to produce simulated data. A virtual environment can go through an infinite number of repetitions and scenarios. The simulated data produced can then be used to train the AI model. This way the AI system can be taught potential real-world conditions, that might otherwise be very rare or still in the testing phase. The second benefit is the ability to plan and test new features. The digital twin should represent reality, but it can produce a view into the future. Are you thinking about investing in a new production line? Are you looking into augmenting your data operations with –machine learning? You can virtually create this world of tomorrow for you and test scenarios. The tests can be tweaked and done as many times as finding the most optimal solution will take. Finally, adding machine learning to any industrial process will make the process more intelligent by getting more accurate data and predictions, and understanding also visual and unstructured data. By adding machine learning into your workflow you don't only open up possibilities to discover previously unseen patterns in your data but also create a single learning-system that can manage complex data" (Alanen 2019).

On a more technical level, about the use of BIM model as a data mine for predictive suggestions in the Operation and Management phase, a team has developed "a BIM-based Data Mining (DM) approach for extracting meaningful laws and patterns, as well as detecting improper records. In this approach, the BIM database is first transformed into a data warehouse. After that, three DM methods are combined to find useful information from the BIM. Specifically, the cluster analysis can find relationships of similarity among records, the outlier detection detects manually input improper data and keeps the database fresh, and the improved pattern mining algorithm finds deeper logic links among records. Particular emphasis is put on introducing the algorithms and how they should be used by building managers. Hence, the value of BIM is increased based on rules, extracted from data of O&M phase that appear irregular and disordered. Validated by an integrated on-site practice in an airport terminal, the proposed DM methods are helpful in prediction, early warning, and decision making, leading to the improvements of resource usage and maintenance efficiency during the O&M phase" (Peng et al. 2017).

On a conceptual level, a relevant research topic is the combination of simulation models and AI. In fact, "…systems inherently exhibit dynamic, goal-seeking, self-preserving, and sometimes evolutionary behaviour, making a system more than simply the sum of its parts, as Meadows emphasised. With such dynamic behaviour, a simulation model may become an inadequate representation of reality over time. This is what is described as the Concept Drift. The Concept Drift refers to modelling a non-stationary problem over time, where a changing context can lead to a mismatch between models and actual problems. In this respect, we pose the following question: what if simulation models can become aware of new information or knowledge in an automated manner? […] The Framework. Key Idea I: Learning to Predict the System Behaviour. Key Idea II: Identify Predictable Influential Variables. Key Idea III: Incremental Learning = Adaptive Behaviour" (Elbattah and Molloy 2018) (Fig. 15).

In sum, we see BIM and AI as two core technologies converging with others into a possibly completely new way of modelling reality: the potential accuracy and fitting of the model, also over time, the opportunity to use past experience to

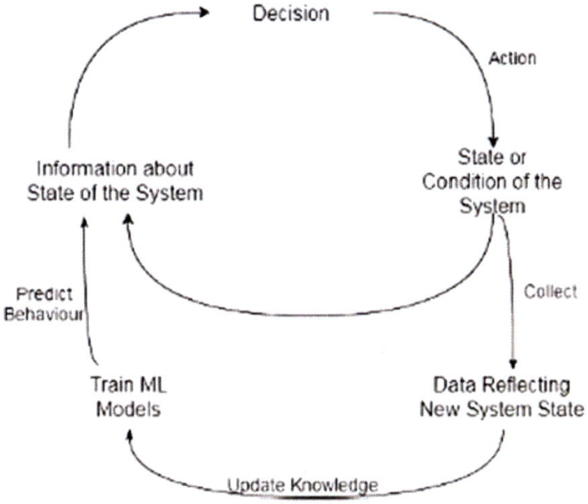

Fig. 15 "Data-driven feedback loop", Fig. 3 in Elbattah and Molloy 2018, https://doi.org/10.1145/3200921.3200933

forecast different cases, the ability to adjust model and predictions in real time and to guide the human counterpart are making BIM and AI way more than a smart drawing tool for architects and engineers, allowing the creation of a comprehensive digital twin of the real world. However, great challenges are ahead, not only as to the technical developments of such techniques, but also as to the organization of the needed knowledge-sharing and workflow-standardization, without which each advancement risks not to benefit from a wider knowledge-base and the validation of such tools would have to be sought each and every time, hampering part of the advantages.

4 AI and BIM: Design Generation and Optimization

Until now we focussed on AI as a means to (semi)-automatize aspects related to BIM projects that could be defined as "technical" as a whole. In fact, even though there are many degrees of freedom and room for discretional choices, even the activity of creating a semantic model cannot be considered the real creative core of architectural design. The uses we've seen tend, in fact, to make good use of data from experience, including technical best practices and best-in-class representation techniques, and devise solutions and lines-of-action based on an immense corpus of shared knowledge within the AEC field, but just from the technical standpoint.

However, the real potential of AI is not necessarily confined to the realm of technical best practices. New experimental work is showing that the same modus operandi of AI could be successfully applied to design optimization and even to its generation. There is a different extent to which AI may be involved into the design process: commercial software packages seem for now limited to most obvious best-practice-based suggestions for plans layout and standard "skins" for building, but some experimental solutions have already gone much further along this path, into more "creative" or "stylistic" aspects of the architectural practice. Also the usage paradigms may vary, ranging from the mere suggestion of alternatives of generated designs based on explicit as well as implicit constraints—among which the architect will chose the most desirable option—to fully automatized design generation algorithms which would directly produce final designs by factoring-in also aesthetical values "learned" from existing design solutions. It will be extremely enticing to see the disruptive impact such a technology may have not just on the work of architects, but even on the idea of architecture and aesthetics.

5 Conclusions

From the foregoing roundup it should emerge clearly that many current BIM-related issues and tools can be highly impacted by the wise use of AI-powered algorithms, while at present it seems such impact is not as relevant as often claimed or believed.

In fact, while specific sub-tasks that are relevant for the construction of a H-BIM model—such as point cloud segmentation, as well as for the checking of a BIM model and for the real-time safety control on the construction yard, among others—can already benefit from the use of AI-powered techniques, it seems that the most advanced prospected uses of AI are still either at the experimentation phase, or at best isolated into niches that still need to be integrated into BIM technology.

Such a *status quo* is probably related to the evolutionary stage of AI itself in the first place, which is far from having reached the level of autonomous intelligence which would be required in order to determine, or at least help, a paradigmatic change over the entire BIM workflow, or at least beyond single repetitive and pre-determined task optimization.

However, some successful proof-of-concept projects already show that more ambitious goals can be hypothesized, making good use of the great amount of data that a series of new technologies are making available and shared, of which BIM sits at the core. The availability of such vast amount of shared data, increasingly difficult to handle by humans, seems indeed the enabling backbone of a purposeful and successful use of AI technologies. For instance, digital survey methods, such as laser scanners and digital photogrammetry, BIM modelling, IoT sensors are all converging technologies producing a large amount of semantically taggable data that are shareable through Cloud-based solutions. What seems yet lacking is—besides the further development of a more general and autonomous kind of AI—a coordinated effort to identify, at least in principle, new streamlined workflow paradigms whereby the whole process of creating and managing the BIM model would eventually constitute a comprehensive tool to bridge between the existing and the wished for, in a self-reinforcing knowledge-based virtuous loop. AI can in fact be usefully applied only when a clear picture of the problems at stake has been clearly defined and stated. Such an effort is what researchers are called in their everyday work, both within private R&D groups and academic institutions, and their joint effort seems needed more than ever for the construction of a global shared standard framework, based on which also the single proprietary solutions and idiosyncratic efforts can find a common ground for the sharing and creation of data and knowledge.

Some extra issues arise at the horizon and start to be discussed in the scientific community. On the one hand, the implications of using AI beyond purely technical tasks, as it is the case for the relationship between AI-powered choices or generative processes and aesthetics, as well as the "affective computing". Another great issue is the shift from research paradigms where a clear distinction is set in principle between causes vs. effects, correlation vs. causation, and where all tools—including statistics—follow the scientific method, to a sometimes obscure "black-box" predictive tool, possibly more effective in providing "correct" results in shorter time, but not able to create explicit knowledge which could be replicated and tested. These are challenges that the scientific community, across disciplines, will have to deal with and find solutions for.

In sum, a great deal of excitement should characterize the research activities related to BIM and AI, as a fundamental step on the path to exponentially grow the body of knowledge needed to assess, renew, plan, design and foresee our built and urban environment in view of human and environmental wellbeing.

References

Alanen P (2019) Digital twin and artificial intelligence: benefits & key learnings, Silo.AI. Aug 5, 2019. https://silo.ai/digital-twin-ai/

Arroyo J, van der Heijde B, Spiessens F, Helsen L (2018) A Python-based toolbox for model predictive control applied to buildings. In: Conference: 5th international high performance building conference At: Purdue University, West Lafayette, Indiana. Project: Flexible Heat and Power, July 2018. https://www.researchgate.net/publication/327136772_A_Python-Based_Toolbox_for_Model_Predictive_Control_Applied_to_Buildings

BIM4EEB Editorial (2019) Renovation process and ontologies: our development, BIM4EEB (Academic consortium), 12 December 2019, Editorial note. https://www.bim4eeb-project.eu/news/renovation-process-and-ontologies-our-development.html

Czerniawski T, Leite F (2019) Semantic segmentation of building point clouds using deep learning: a method for creating training data using BIM to point cloud label transfer. In: ASCE International conference on computing in civil engineering 2019, conference paper. https://ascelibrary.org/doi/10.1061/9780784482421.052, https://doi.org/10.1061/9780784482421)

Day M (2019) Augmented thinking—how AI is coming to AEC, AECmagazine, 18 Nov 2019, On-line Magazine. https://aecmag.com/component/content/article/59-features/1904-augmented-thinking-how-ai-is-coming-to-aec

Drgoňa J, Picard D, Kvasnica M, Helsena L (2018) Approximate model predictive building control via machine learning. Appl Energy 218:199–216. https://doi.org/10.1016/j.apenergy.2018.02.156

Intel Editorial (2015) Controlling office temperature with building robotics. An environmental management solution empowers building occupants, Intel Solution Brief, 2015. https://www.intel.com/content/dam/www/public/us/en/documents/white-papers/iot-smart-office-building-robotics-brief.pdf

Elbattah M, Molloy O (2018) ML-Aided Simulation: A Conceptual Framework for Integrating Simulation Models with Machine Learning. In: SIGSIM-PADS '18, proceedings of the 2018 ACM SIGSIM conference on principles of advanced discrete simulation, May 2018 pp 33–36. https://doi.org/10.1145/3200921.3200933

Elnagar S, Yoon V, Thomas M (2020) An automatic ontology generation framework with an organizational perspective. In: Proceedings of the 53rd Hawaii international conference on system sciences 2020, 6 Jan 2020. http://hdl.handle.net/10125/64339

Isokane T, Okura F, Ide A, Matsushita Y, Yagi Y (2018) Probabilistic plant modeling via multi-view image-to-image translation. arXiv:1804.09404

Khaleel Seeaed M, Hamdan A-H (2019) BIMification of stone walls for maintenance management by utilizing Linked Data, Conference: 31.Forum Bauinformatik, Berlin, November 2019, Conference paper, https://www.researchgate.net/publication/337104176_BIMification_of_stone_walls_for_maintenance_management_by_utilizing_Linked_Data

Kolyvakis P, Kalousis A, Kiritsis D (2018) DeepAlignment: unsupervised ontology matching with refined word vectors. In: Proceedings of the 2018 Conference of the North American Chapter of the Association for Computational Linguistics: Human Language Technologies, vol 1 (Long Papers), June 2018. https://doi.org/10.18653/v1/n18-1072, https://www.aclweb.org/anthology/N18-1072/

Maietti F, Di Giulio R, Piaia E, Medici M, Ferrari F (2018) Enhancing heritage fruition through 3D semantic modelling and digital tools: the inception project, florence Heri-Tech—the future of heritage science and technologies. IOP confernce series: materials science and engineering vol 364, p 012089. https://doi.org/10.1088/1757-899x/364/1/012089

MIT affecting computing group, project "Predicting students' wellbeing from physiology, phone, mobility, and behavioral data", MIT affecting computing group, research group site. https://www.media.mit.edu/projects/predicting-students-wellbeing-from-physiology-phone-mobility-and-behavioral-data/overview/

Musella C, Serra M, Menna C, Asprone D (2019) BIM & AI: advanced technologies for the digitalisation of seismic damages in masonry buildings. In: Conference: international fib symposium on conceptual design of structures, Madrid, Sept 2019. https://www.researchgate.net/publication/336085678_BIM_AI_advanced_technologies_for_the_digitalisation_of_seismic_damages_in_masonry_buildings

Palmirani M, Bincoletto G, Leone V, Sapienza S, Sovrano F (2019) PrOnto ontology refinement through open knowledge extraction. In: Araszkiewicz M, Rodríguez-Doncel V, the thirty-second annual conference legal knowledge and information systems: JURIX 2019, vol 322 of frontiers in artificial intelligence and applications, 272 pages. IOS Press. https://doi.org/10.3233/faia190326. ISBN 1643680498, 9781643680491

Pauwels P (2019) Linked data and the semantic Web: the Basics, 17 June 2019, Slides for LDAC Summer School. http://linkedbuildingdata.net/ldac2019/summerschool/files/02_Pauwels_Linked_Data_and_SWT_Basics.pdf

Peng Y, Lin J-R, Zhang J-P, Hu Z-Z (2017) A hybrid data mining approach on BIM-based building operation and maintenance. Build Environ 126:483–495. https://doi.org/10.1016/j.buildenv.2017.09.030

Qi C-R, Su H, Mo K, Guibas L-J (2017). PointNet: deep learning on point sets for 3D classification and segmentation.In: Conference on Computer Vision and Pattern Recognition (CVPR) 2017. http://stanford.edu/~rqi/pointnet/, https://arxiv.org/abs/1612.00593v2

Rasmussen M-H, Pauwels P, Hviid C-A, Karlshøj J (2017) Proposing a central AEC ontology that allows for domain specific extensions. In: Lean and Computing in Construction Congress (LC3): Volume I Ð Proceedings of the Joint Conference on Computing in Construction (JC3), July 4–7, 2017, Heraklion, Greece, pp 237–244. http://itc.scix.net/paper/lc3-2017-153, https://doi.org/10.24928/jc3-2017/0153

Runceanu L-S, Haala N (2018) Indoor mesh classification for BIM. Int Arch Photogramm Remote Sens Spatial Inf Sci XLII-4:535–539. https://doi.org/10.5194/isprs-archives-XLII-4-535-2018

Scherer R-J (2017) BIM ification eine neue Herausforderung an die Bauinformatik, Prof. Raimar Scherer FBI, Dresden, Dresden, Slides. http://docplayer.org/75889235-Bimification-eine-neue-herausforderung-an-die-bauinformatik.html

Zlatanova S, Beetz J, Goos J et al (2014) 3D spatial infrastructure for the Port of Rotterdam, GISt Report No. 67 TU Delft. https://www.researchgate.net/publication/269688807_3D_Spatial_Infrastructure_for_the_Port_of_Rotterdam